Printed in the United States of America.

INTEGR8D.org

ON PROTRACTED WAR

Table of Contents

Foreword

The Systems Logic of Liberation

When Mao Zedong delivered these lectures in Yan'an in 1938, the world was witnessing the collapse of established orders. Conventional military wisdom held that China, a semi-colonial nation with inferior weaponry, would inevitably fall to the imperial machinery of Japan. Mao, however, offered more than a counter-strategy. He articulated a theory of systemic transformation.

Read today through the lens of the Systems Thinking Framework (STF), *On Protracted War* reveals a mode of analysis strikingly similar to that of Ibn Khaldun. Both thinkers, in different contexts, sought to decode the lifecycle of power, the fragility of overextended systems, and the generative capacity of collective solidarity. STF does not collapse or conflate their differences. Instead, STF offers a clarifying analytical lens through which shared structural patterns become visible across time.

I write this foreword to situate Mao's strategic thought within a broader civilizational context, drawing parallels between his analysis of the Sino-Japanese War and Ibn Khaldūn's sociology of rise and decline. These parallels are illuminated through the abstraction $S(X) = f(W, E)/t$, where systems produce outcomes (X) through the interaction of Work (W), Energy (E), and Time (t). Here, time is not merely a divisor but a condition of realization. In this logic, time is a non-compressible variable; attempts to reduce t toward zero without a corresponding reduction in the system's complexity lead to the collapse of the possibility of actualization.

In the STF, one guiding principle, derived from Ibn Khadlun's ideas, holds that excess produces its opposite. Systems optimized for a singular objective, when extended beyond their adaptive capacity, generate outcomes that undermine their original purpose. Ibn Khaldūn described this trajectory in terms of the movement from ʿumrān (the field of human association) to ḥaḍāra (its complex and surplus-driven configuration) wherein luxury erodes social cohesion (ʿaṣabiyya) and weakens the state from within.

Mao's analysis of Japan reflects a comparable structural insight. He identifies Japan as a "strong" imperialist power whose strength rests on foundations he characterizes as "retrogressive and barbarous." In STF terms, Japan's war system is highly optimized for rapid extraction and territorial expansion, forms of efficient work, but in ways that, under conditions of prolonged resistance, tend toward systemic fragility. This over-optimization neglects contributory systems such as morale, international legitimacy, and logistical sustainability. As Mao observes, these neglected dimensions form "the principal basis for its inevitable defeat."

The flaw in Japan's strategy is not simply moral but temporal. Its pursuit of "quick victory" reflects a failure to grasp time as a constitutive variable. It attempts to compress duration in order to maximize immediate output, hence immediate gain, like most self-interested actors do. However, systems cannot actualize outcomes without temporal extension, for time is the invisible fabric that connects the whole. By underestimating the time required for political consolidation and the energy required to sustain occupation, Japan's apparent efficiency generates the very conditions of its exhaustion. The empire becomes captive to its own momentum, unable to recalibrate without undermining the logic of its expansion and immediate reward.

If Japan represents the over-optimized system, China represents the reconstitution of systemic energy. In Ibn Khaldūn's thought, *ʿaṣabiyya* is the driving force that enables collective action. Within STF, this corresponds to Energy (E): the structured capacity of a system to mobilize and sustain coordinated work.

Mao's central claim is that victory depends not on weapons, but on people. "The decisive factor is man, not things." This is a systems statement. Weapons are stored work; people are the source of renewable energy. Mao's emphasis on political mobilization is therefore foundational. Without the activation of social *energy*, the system cannot sustain the work required for victory.

In this framework, China's success depends on aligning specific forms of work—guerrilla warfare, mobile operations, political organization— with a corresponding energy base: national unity, legitimacy, and shared purpose. While energy ultimately enters any system from outside its immediate configuration, in social systems this "externality" is internalized as identity, dignity, and collective will. Mao captures this transformation vividly: mobilizing the population creates "a vast ocean", a collective or a whole, in which the enemy is submerged. Latent energy becomes active force.

The convergence between Mao, Ibn Khaldūn, and the STF is most evident in their treatment of time. Time is not a passive container but a dynamic condition within which work accumulates effects and energy encounters limits. Mao's concept of *protracted* war is, at its core, a strategic reconfiguration of *time*. He divides the conflict into three stages: Strategic Defense, Strategic Stalemate, and Strategic Counteroffensive. These stages describe a lifecycle of systemic transformation:

- Defense: Time is acquired and energy preserved. Space is traded to prevent collapse.
- Stalemate: Energy is rebalanced. The overextended system begins to exhaust itself, while dispersed forms of work accumulate strength.
- Counteroffensive: The balance shifts. Sustained work, supported by regenerated energy, converts temporal endurance into strategic advantage.

Here, strength and weakness are more than fixed attributes. They are functions of *time*. As Mao notes, unfavorable and favorable conditions evolve with the prolongation of war. Systems endure not by resisting change, but by dynamically adapting before imbalance becomes rupture, for change is inevitable per STF principles. China adapts; Japan, constrained by its imperial logic, cannot. This dynamic mirrors Ibn Khaldūn's concept of *tadāwul*, the gradual alternation through which transformation unfolds before becoming visible.

These insights remain relevant because the structural dynamics they describe persist in new forms. Contemporary systems of economic extraction, surveillance, and military projection often exhibit similar patterns: they optimize for control, efficiency, or scale while neglecting legitimacy, regeneration, and temporal sustainability. Such imbalances do not immediately produce collapse, but they introduce constraints that, over time, reshape outcomes.

On Protracted War is therefore more than a military text. It demonstrates how a non-determinant system can overcome a stronger one by aligning work, energy, and time. It shows that determinant systems can be disrupted when their energy sources are strained, that contributory systems can shift overall balance when amplified, and that time becomes advantageous only when sustained work is matched by regenerative energy.

Within the STF, rights are not abstractions but outcomes of functioning systems. Mao illustrates this in practice: national *liberation* is not granted but produced through coordinated effort over time. The decline of an oppressive system is not metaphysical inevitability but the result of structural imbalance. When a system persistently ignores the forces upon which it depends—above all, the people who do the work—it generates the conditions of its own reversal.

At the same time, it is important to maintain analytical precision. This comparison does not collapse Mao into Ibn Khaldūn, nor does it recast Ibn Khaldūn as a precursor to modern revolutionary thought. They belong to distinct intellectual traditions and historical contexts. What unites them is not ideology, but a shared commitment to identifying underlying principles that govern social transformation independent of culture and history.

The STF does not claim to replace their individual perspectives. It provides a language through which their insights can be read alongside one another without conflation. Each thinker isolates structural relationships between strength and weakness, cohesion and fragmentation, duration and change, and uses them to explain and anticipate systemic outcomes. That similar patterns emerge across such different contexts suggests not a shared doctrine, but the recurrence of constraints inherent to human association—the presence of social systems.

One final point of convergence deserves emphasis: the primacy of work (*'amal*). For Ibn Khaldūn, work is the generative activity through which value comes into being. For Mao, people, the agents of work, are decisive, while material instruments remain derivative. In both cases, value is rooted in human activity rather than abstract accumulation. Capital, in this light, represents stored work; it cannot substitute for the ongoing processes that sustain a system.

This is not an argument for uniformity, but a recognition of limits. Systems that privilege accumulation over generation, or efficiency over balance, tend toward instability. By making work, energy, and time explicit variables, the STF enables a more precise diagnosis of these tendencies and the conditions under which they can be corrected.

As you read this text, look beyond its military vocabulary. Attend instead to its structure: the decomposition of conflict into interacting variables, the recognition that no system operates in isolation, and the insistence that outcomes emerge from relationships rather than isolated forces. Mao's analysis connects the war in China to global dynamics, domestic conditions, and shifting alignments. It is, in this sense, a form of systems thinking.

The enduring relevance of *On Protracted War* lies in this deeper architecture. Across contexts, whether in struggles against domination, inequity, or ecological crisis, the same lesson recurs: outcomes do not belong to those with the greatest force, but to those who can align work, energy, and time within a coherent and adaptive system.

One final clarification is necessary, particularly in light of modern economic debates, to avoid a common misreading of Ibn Khaldun. His emphasis on work should not be mistaken for an anticipation of modern socialism or communism. Although he places work at the center of value creation, he does not center the *worker* as a class or social category in the manner of modern ideologies. His analysis is not concerned with elevating a social group, but with identifying the fundamental process through which human existence within society is sustained.

In *al-Muqaddima*, Ibn Khaldūn developed and nuanced vocabulary like *rizq*, *kasb*, *sa`y*, and *tahsil*. But the universal element that sustains life is work (*`amal*). What humans require for survival is not available without work, and no individual can secure these necessities in isolation. Human existence, therefore, is inseparable from cooperation (*ta`āwun*), and cooperation itself is structured around the performance and distribution of *work*. This is not a metaphysical claim about existence in the abstract, but a social-functional one: within *`umrān*, human life is sustained through organized activity. *Work*, in this sense, is not one variable among others; it is the condition through which all others become operative.

This distinction is crucial when compared to modern frameworks. Contemporary economic systems tend to oscillate between two poles: those that prioritize capital and those that prioritize workers. Capital-centered systems organize institutions around accumulation, often treating capital as the primary driver of production and value. Worker-centered systems, by contrast, elevate labor as a social category, seeking to structure economic life around its interests and claims. Each captures part of the reality: without accumulated resources, coordination at scale becomes difficult; without human effort, no value can be generated. However, both remain partial because they elevate elements within the system rather than the generative process that underlies them.

Ibn Khaldūn's position is different in kind. He centers neither capital nor the worker, but *work* itself as the defining process through which value is produced and life is sustained. From this perspective, capital does not generate value independently. Capital preserves the results of prior work. Money, whether in the form of gold (*dīnār*) or its equivalents, functions as a store of realized work, not as a self-generating source of value. To treat it otherwise is to detach value from its generative basis.

This principle is applied with striking consistency across his analysis. When Ibn Khaldūn distinguishes between the human being in potential (*ādamī bi al-quwwa*) and in actuality (*ādamī bi al-fiʿl*), he marks the transition at separation (*fiṣāl, infiṣāl*). In his discussion of midwifery, Ibn Khaldun explains that midwives are more knowledgeable than physicians about the ailments of infants during the period of suckling, up to *fiṣāl*— the stage when the child is weaned and separated from dependence on the mother's milk. He then clarifies why: during this entire phase, the infant's body is only a "human body in potential" (*biʾl-quwwa*), because it does not yet sustain itself through its own processes (work). It lives off the mother's work—her acquisition, digestion, and transformation of food into nourishment. Only after *fiṣāl*, when the child begins to eat and process more food independently, does it become a "human body in actuality" (*biʾl-fiʿl*), functioning as its own self-operating system. This transition reflects a broader principle: realized human existence is marked by the capacity to perform sustaining activity. As this capacity develops, so too does the individual's ability to secure livelihood (*rizq*) and generate value through acquisition (*kasb*). The distinction reflects a consistent view that existence, within human society (and perhaps all biological beings), is inseparable from the performance of *work*.

A work-centered framework thus differs fundamentally from both capital-centered and worker-centered systems. It does not privilege accumulated wealth as the organizing principle, nor does it reify a social class as the locus of value. Instead, it directs attention to the processes through which work is generated, sustained, and distributed across a system. Capital remains necessary, but derivative; human agents remain essential, but are not reducible to fixed categories. The central analytical question becomes whether a system enables the continuous production of meaningful work and aligns its rewards with that generative activity.

In the language of the Systems Thinking Framework, this restores primacy to Work (W) while preserving its relation to Energy (E) and Time (t). Systems that center capital risk detaching stored work from its generative source, allowing accumulation to proceed without corresponding production. Systems that center workers risk overlooking the structural conditions that make sustained and transferable work possible. A work-centered system, by contrast, evaluates both capital and human participation in terms of their contribution to ongoing processes of production, renewal, and balance (transformation).

Ibn Khaldūn does not recognize "workers" as a discrete or foundational social class in the manner of modern economic thought. This is not because work is marginal in his analysis, but because it is universal and constitutive. Work (ʿamal) is the activity through which human life within ʿumrān is sustained and realized. It is not confined to a particular segment of society, nor does it serve as the basis for a unified social identity. While Ibn Khaldūn does distinguish between different forms of livelihood and occupational roles, these distinctions do not coalesce into a singular category of "workers" as an analytical or political class. The emphasis remains on the processes of activity and cooperation that make social existence possible, rather than on the elevation of one group within it.

From this follows a second point. Ibn Khaldūn does not treat the division between producers and consumers, or between labor and capital, as primary or structuring oppositions. Although his analysis clearly recognizes differences in function, surplus, and modes of acquisition, these distinctions arise within a broader system organized around the generation and distribution of work. In this sense, what later frameworks describe as "capital" can be understood as the accumulated result of prior activity, while consumption remains dependent on ongoing processes of production. The separation of these elements into discrete and opposing categories reflects a later analytical development rather than a foundational feature of his thought. A work-centered reading thus shifts attention away from fixed binaries and toward the continuous processes through which value is created, sustained, and circulated within society.

From this perspective, certain features of modern economic life reveal structural tension. When capital generates returns without a clear connection to productive activity, the relationship between work and value becomes attenuated. The result is more than distributive imbalance; it points to systemic fragility: if value can expand without work, it can also contract independently of it, undermining the processes that sustain livelihoods. In Ibn Khaldūn's terms, such a condition approaches *fasād*—a corruption in the relationship between effort and reward that weakens the coherence of the system itself.

For this reason, few contemporary state systems are explicitly organized around a primary commitment to work—as distinct from capital or labor as a social category—as the foundational principle of value and policy. Existing models tend to privilege either accumulation or distribution, rather than the generative processes that make both possible. Ibn Khaldūn's analysis points toward a different orientation: one in which work is treated as the basis of value, capital as its stored form, and social organization as the structure that enables their alignment.

Seen in this light, his contribution is not a precursor to modern ideological divisions, but a reframing that precedes them. It invites us to move beyond the binary of capital versus labor and to ask a more fundamental question: what forms of work sustain human existence, and how should systems be organized to support and reward them? In raising this question, Ibn Khaldūn offers not a doctrine, but a principle—one grounded in the observable conditions of human life, and one that remains as relevant today as in the world he sought to understand.

This work-centered reading of Ibn Khaldun also clarifies, in a deeper way, the practical philosophy of Mao Zedong. When Mao insists that "the decisive factor is man, not things," he is not merely privileging people over material resources; he is identifying the same underlying principle: that only living systems capable of performing work can generate and sustain outcomes. Weapons, capital, and infrastructure are all forms of stored work, but without active human engagement, they remain inert. In this sense, Mao's emphasis on mobilization is not ideological alone—it is systemic. A population that does not act cannot produce victory, just as a body that does not perform its own sustaining processes remains in a state of potential.

Seen through this lens, Mao's strategy of protracted war becomes a large-scale application of the same logic Ibn Khaldun identifies at the level of the individual. Just as the infant becomes fully actual only when it begins to perform the work necessary to sustain itself, a society becomes historically effective only when it activates its own internal capacities for coordinated action. Liberation, therefore, is not granted by external conditions or superior force; it emerges when a system transitions from dependence to self-sustaining activity. In both thinkers, the movement from potential to actuality marks the moment when a system—whether a human body or a political community—begins to do its own work and, in doing so, becomes capable of shaping its own outcomes over time.

Ahmed E. Souaiaia, PhD; March 16, 2026

ON PROTRACTED WAR

(May 1938)

This is the text of a speech delivered by Mao Tse-Tung before the Yan'an Association for the Study of the War of Resistance Against Japan, delivered between May 26 and June 3, 1938.

I. The Raising of the Questions

(1) The first anniversary of the great War of Resistance Against Japan—the seventh of July—draws near. For nearly a year now, the entire nation has united its strength, persisted in resistance, upheld the united front, and waged a heroic struggle against the enemy. This war is unprecedented in the history of the East and will prove to be of monumental significance in world history; it commands the concern of peoples everywhere. Every Chinese citizen, suffering the calamities of war and striving for the survival of our nation, longs day by day for victory. Yet what, precisely, will be the course of this war? Can victory be attained, or not? Can it be won swiftly, or not? Many speak of a protracted war—but why must it be protracted? How is such a war to be conducted? Many speak of ultimate victory—but why is ultimate victory assured? How is it to be secured? These questions have not been resolved by all; indeed, the majority have yet to arrive at satisfactory answers. Consequently, proponents of defeatist "national subjugation theory" have emerged, declaring to the people: "China will perish; ultimate victory will not belong to China." Meanwhile, certain impetuous friends have also come forth, asserting: "China can achieve victory very soon; there is no need to exert tremendous effort." Are these arguments correct? We have consistently maintained that they are not. Yet what we have said has not been understood by the majority. This is due in part to the insufficiency of our propaganda and explanatory work, and in part to the fact that the objective course of events has not yet fully revealed its inherent nature nor clearly presented its countenance before the people, leaving them unable to discern the overall trend and prospects, and thus unable to determine their comprehensive policy and approach. Now, however, the experience of ten months of resistance suffices to refute the baseless theory of national subjugation and to persuade our impetuous friends of the fallacy of the theory of quick victory. Under these circumstances, many have called for a summative explanation. Particularly regarding the protracted war, there exist not only opposing views from the theories of subjugation and quick victory, but also superficial and vacuous understandings. A formula such as—"Since the Lugou Bridge Incident [1], four hundred million people have exerted themselves as one; ultimate victory belongs to China"—circulates widely among the populace. This formula is correct, yet it requires substantiation. The perseverance of the War of Resistance and of the united front is attributable to numerous factors: political parties across the nation, from the Communist Party to the Kuomintang; the entire people, from workers and peasants to the bourgeoisie; all armed forces, from the main forces to guerrilla units; international actors, from socialist states to justice-loving peoples of all nations; and elements within the enemy country itself, from certain anti-war civilians to soldiers at the front opposing the conflict. In sum, all these factors have contributed, each to varying degrees, to our resistance. Every person of conscience ought to express respect toward them. We Communists, together with other resistance parties and the entire Chinese people, have but one direction: to strive to unite all forces and defeat the vile Japanese aggressors. July 1 of this year marks the seventeenth anniversary of the founding of the Communist Party of China. To enable every Communist Party member to exert still greater and more effective efforts in the War of Resistance, it is essential to undertake a focused study of the protracted war. Accordingly, my address will examine the protracted war. I intend to address all questions pertinent to this subject; yet I cannot address everything, for not all matters can be fully expounded within a single speech.

(2) The experience of the ten months of resistance has demonstrated the incorrectness of two viewpoints: one, the theory that China is bound to perish; the other, the theory that China can achieve swift victory. The former fosters a tendency toward compromise; the latter, a tendency toward underestimating the enemy. Their method of analyzing problems is subjective and one-sided—in a word, unscientific.

(3) Prior to the War of Resistance, numerous arguments advancing the theory of national subjugation prevailed. For example: "China's weaponry is inferior to that of others; should we resist, we are certain to be defeated." Or: "If we resist, we shall surely become another Abyssinia [2]." After the outbreak of resistance, overt expressions of the subjugation theory have ceased, yet covert adherence to it persists—and is, in fact, widespread. For instance, sentiments favoring compromise ebb and flow; the basis upon which advocates of compromise rest their position is the assertion that "further resistance will surely lead to perdition" [3]. A student writing from

Hunan reported: "In the countryside, everything feels difficult. Working alone in propaganda, I can only seek conversations with people wherever I go. My interlocutors are not ignorant fools; they understand something, and they take interest in what I say. But when I encounter certain relatives of mine, they invariably declare: 'China cannot win; it will perish.' How vexing! Fortunately, they do not propagate these views; otherwise, the situation would be truly dire. Naturally, the peasantry places greater trust in them!" Such proponents of the "China-is-bound-to-perish" theory constitute the social foundation of the tendency toward compromise. Such individuals exist throughout China; consequently, the issue of compromise, which may arise at any moment within the anti-Japanese front, will likely persist until the war's conclusion. At this critical juncture, with the loss of Xuzhou and the tension surrounding Wuhan, it would not be unhelpful to deliver a thorough refutation of this defeatist theory.

(4) Over the ten months of resistance, various manifestations of impetuous thinking have also emerged. For example, at the outset of the war, many exhibited an unfounded optimism, underestimating Japan to the extent of believing it incapable of advancing into Shanxi. Some disparaged the strategic significance of guerrilla warfare within the War of Resistance, expressing skepticism toward the proposition that "strategically, mobile warfare is primary and guerrilla warfare supplementary; tactically, guerrilla warfare is primary and mobile warfare supplementary." They disapproved of the Eighth Route Army's strategic guideline: "Guerrilla warfare is fundamental, but mobile warfare under favorable conditions must not be relaxed," dismissing it as a "mechanical" viewpoint [4]. During the Shanghai campaign, some asserted: "If we can hold out for three months, the international situation will certainly change; the Soviet Union will certainly intervene, and the war will be resolved." Such views placed the prospects of resistance primarily upon foreign assistance [5]. Following the victory at Taierzhuang [6], some argued that the Xuzhou campaign [7] should constitute a "quasi-decisive battle," insisting that the previous guideline of protracted war ought to be revised. They proclaimed: "This battle represents the enemy's final struggle"; "Once we prevail, the Japanese militarists will have lost their moral footing and can only await final judgment" [8]. The victory at Pingxingguan intoxicated some; the subsequent victory at Taierzhuang intoxicated still more. Consequently, whether the enemy would advance on Wuhan became a matter of doubt. Many believed: "Not necessarily"; many insisted: "Absolutely not." Such doubts extend to all major questions. For instance: Is our anti-Japanese strength sufficient? One might answer affirmatively, reasoning that our current strength has already halted the enemy's advance—why, then, increase it further? For instance: Is the slogan of consolidating and expanding the anti-Japanese national united front still correct? One might answer negatively, reasoning that the present state of the united front suffices to repel the enemy—what need, then, for consolidation or expansion? Similarly, one might answer negatively regarding whether international diplomacy and propaganda efforts should be intensified; whether reforms of military and political institutions, development of mass movements, rigorous implementation of national defense education, suppression of traitors and Trotskyites [9], development of military industry, and improvement of living standards ought to be earnestly pursued; or whether the slogans of defending Wuhan, defending Guangzhou, defending the Northwest, and vigorously developing guerrilla warfare behind enemy lines remain valid. Indeed, whenever the military situation improves somewhat, certain individuals prepare to intensify friction between the Kuomintang and the Communist Party, shifting their focus from external to internal affairs. This pattern recurs after nearly every significant victory or temporary pause in enemy offensives. All of the foregoing we designate as political and military myopia. Such arguments may sound plausible, yet in reality they are baseless, specious, and empty. Eliminating such empty talk should prove beneficial to the conduct of a victorious War of Resistance.

(5) Thus the questions are: Will China perish? Answer: It will not perish; ultimate victory belongs to China. Can China achieve swift victory? Answer: It cannot; the War of Resistance Against Japan is a protracted war.

(6) The essential arguments concerning these questions were generally outlined by us more than two years ago. As early as July 16, 1936—five months prior to the Xi'an Incident and twelve months before the Lugou Bridge Incident—in my conversation with the American journalist Mr. Edgar Snow, I already provided a general assessment of the situation in the Sino-Japanese conflict and proposed various guidelines for securing victory. For reference, I quote several excerpts below:

Q: Under what conditions can China defeat and eliminate the power of Japanese imperialism?

A: Three conditions are required: First, the completion of China's anti-Japanese united front; second, the completion of an international anti-

Japanese united front; third, the rise of revolutionary movements among the Japanese people and the peoples of Japan's colonies. From the standpoint of the Chinese people, the great unity of the Chinese people is the primary condition among these three.

Q: How long, in your view, will this war last?

A: That depends on the strength of China's anti-Japanese united front and on many other decisive factors concerning China and Japan. That is to say, besides primarily considering China's own strength, the assistance China receives internationally and the revolutionary assistance arising within Japan are also highly relevant. If China's anti-Japanese united front develops robustly and is effectively organized both horizontally and vertically; if governments and peoples of other nations, recognizing that Japanese imperialism threatens their own interests, provide China with necessary assistance; and if revolution in Japan arises swiftly—then this war will conclude rapidly, and China will achieve swift victory. If these conditions cannot be realized quickly, the war will be prolonged. Yet the outcome remains the same: Japan will inevitably be defeated, and China will inevitably triumph. The sacrifice, however, will be great, and a very arduous period must be endured.

Q: Politically and militarily, how do you envision the development of this war's prospects?

A: Japan's continental policy has already been determined. Those who imagine that compromising with Japan and sacrificing additional Chinese territory and sovereignty could halt Japanese aggression are indulging in mere fantasy. We know with certainty that even the lower Yangtze region and southern ports are already encompassed within Japanese imperialism's continental policy. Moreover, Japan seeks to occupy the Philippines, Siam, Vietnam, the Malay Peninsula, and the Dutch East Indies, thereby severing foreign connections with China and monopolizing the Southwest Pacific. This constitutes Japan's maritime policy. During such a period, China will undoubtedly face extreme difficulties. Yet the majority of Chinese people believe these difficulties can be overcome; only the wealthy in major commercial ports are defeatists, for they fear loss of property. Many suppose that once Japan blockades China's coast, China cannot continue the war. This is nonsense. To refute them, we need only cite the military history of the Red Army. In the War of Resistance, China enjoys advantages far superior to the position of the Red Army during the civil war. China is a vast country; even if Japan occupies regions containing one to two hundred million people, we remain far from defeat. We still possess substantial strength to fight Japan, while Japan, throughout the war, must constantly maintain defensive operations in its rear. China's economic disunity and imbalance, paradoxically, prove advantageous for the War of Resistance. For example, severing Shanghai from other parts of China would inflict far less damage on China than severing New York from the rest of the United States would inflict on America. Even if Japan blockades China's coast, it cannot blockade China's northwest, southwest, and western regions. Thus, the crux of the matter remains: the entire Chinese people must unite and establish a nationwide, unified anti-Japanese front. This is a proposition we advanced long ago.

Q: If the war is prolonged and Japan is not completely defeated, can the Communist Party agree to negotiations and recognize Japanese rule over the Northeast?
A: No. The Communist Party of China, like the entire Chinese people, will not permit Japan to retain a single inch of Chinese territory.

Q: In your view, what should be the principal strategic guideline for this war of liberation?

A: Our strategic guideline should be to employ our main forces to operate along extensive, fluid fronts. For the Chinese military to prevail, it must conduct highly mobile warfare across vast battlefields—advancing and retreating rapidly, concentrating and dispersing swiftly. This constitutes large-scale mobile warfare, not static positional warfare reliant on deep trenches, layered fortifications, and exclusive dependence on defensive works. This does not mean abandoning all strategically important military locations; where advantageous, positional warfare should be employed for such points. However, the strategic guideline governing the overall situation must necessarily be mobile warfare. Positional warfare, though also necessary, belongs to the secondary, auxiliary category. Geographically, given the vastness of the battlefield, it is feasible for us to conduct the most effective mobile warfare. Japanese forces, encountering our vigorous mobile operations, will necessarily proceed with caution. Their war machine is cumbersome, slow in movement, and limited in efficacy. If we concentrate our forces to wage a war of attrition within a narrow positional front, we would forfeit our advantageous geographical and economic-organizational conditions and repeat the error of Abyssinia. In the early phase of the war, we must avoid all major decisive engagements; instead, we should gradually undermine the enemy's morale and combat effectiveness through mobile warfare.

In addition to deploying trained troops for mobile warfare, we must organize numerous guerrilla units among the peasantry. It should be noted that the Anti-Japanese Volunteer Armies in the three northeastern provinces represent merely a small fraction of the latent resistance potential that can be mobilized among China's peasantry. China's peasants possess immense latent strength;

if properly organized and directed, they can keep Japanese forces occupied twenty-four hours a day, wearing them down through constant harassment. It must be remembered that this war is being fought on Chinese soil—that is to say, Japanese forces will be entirely surrounded by a hostile Chinese population; they will be compelled to transport their own military supplies and guard them personally; they must deploy substantial forces to protect lines of communication, ever vigilant against ambushes; and furthermore, they must maintain a significant portion of their strength stationed in Manchuria and the Japanese homeland.

In the course of the war, China will be able to capture many Japanese soldiers and seize considerable quantities of weapons and ammunition to arm itself; simultaneously, by securing foreign assistance, the equipment of Chinese forces can be gradually strengthened. Consequently, China will be able to engage in positional warfare during the later stages of the conflict and launch positional attacks against Japanese-occupied territories. Thus, under the prolonged attrition of China's resistance, Japan's economy will approach collapse; amid countless military engagements, its morale will inevitably deteriorate. On China's side, the latent potential for resistance will surge ever higher day by day, as great numbers of revolutionary masses continuously pour to the front to fight for freedom. All these factors, combined with others, will enable us to launch a final, decisive assault against Japan's fortified positions and base areas in occupied China, and to expel the Japanese aggressors from our country. The experience of ten months of resistance has confirmed the correctness of the foregoing arguments, and will continue to confirm them in the future.

(7) More than a month after the Lugou Bridge Incident—on August 25, 1937—the Central Committee of the Communist Party of China clearly stated in its *Decision on the Current Situation and the Tasks of the Party*:

The challenge at Lugou Bridge and the occupation of Beiping and Tianjin merely mark the beginning of the Japanese aggressors' large-scale offensive against China proper. The enemy has already initiated nationwide wartime mobilization. Their so-called propaganda of "not seeking expansion" is nothing but a smokescreen to conceal their offensive.

The resistance at Lugou Bridge on July 7 has become the starting point of China's nationwide War of Resistance.

China's political situation has thereby entered a new stage: the stage of implementing resistance. The preparatory stage of resistance has passed.

The central task of this stage is: to mobilize all forces to secure victory in the War of Resistance.

The central key to achieving victory in the War of Resistance lies in developing the already-commenced resistance into a comprehensive, nationwide, people's war. Only such a comprehensive, nationwide, people's war can secure the ultimate victory of the resistance.

Owing to serious weaknesses that still exist in the current resistance, many adverse developments may occur in the course of the war going forward: setbacks, retreats, internal fragmentation, defections, and temporary or partial compromises. Therefore, it must be recognized that this resistance is an arduous, protracted war. Yet we believe that the resistance already launched will, through the efforts of our Party and the entire Chinese people, overcome all obstacles and continue to advance and develop.

The experience of ten months of resistance has likewise confirmed the correctness of the foregoing arguments, and will continue to confirm them in the future.

(8) Idealist and mechanistic tendencies in the analysis of war constitute the epistemological root of all erroneous viewpoints. Their method of examining problems is subjective and one-sided: either they indulge in purely subjective speculation without basis; or they take a single aspect or temporary manifestation of a problem and, equally subjectively, exaggerate it into the whole. However, erroneous viewpoints may be divided into two categories: one comprises fundamental errors, characterized by consistency, which are difficult to correct; the other comprises incidental errors, characterized by transience, which are relatively easy to correct. Yet since both are errors, both require correction. Therefore, only by opposing idealist and mechanistic tendencies in the analysis of war, and by adopting an objective and comprehensive viewpoint to examine war, can correct conclusions regarding war be reached.

II. The Basis of the Problem

(9) Why is the War of Resistance Against Japan a protracted war? Why is ultimate victory destined to belong to China? Wherein lies the basis for these conclusions?

The Sino-Japanese War is no ordinary war; it is a war of life-and-death struggle waged in the 1930s between semi-colonial, semi-feudal China and imperialist Japan. The entirety of the problem finds its basis herein. Speaking specifically, the two belligerents exhibit the following mutually opposing characteristics.

(10) Regarding Japan:

First, it is a powerful imperialist nation. Its military strength, economic capacity, and political organizational power rank first in the East and place it among the five or six foremost imperialist powers in the world. This constitutes the fundamental condition for Japan's war of aggression; the inevitability of war and China's inability to achieve swift victory rest precisely upon Japan's imperialist system and its formidable military, economic, and political-organizational strength.

However, second, owing to the imperialist character of Japan's socio-economic structure, its war assumes an imperialist character: it is retrogressive and barbarous. By the 1930s, Japanese imperialism, beset by internal and external contradictions, found itself compelled not only to embark upon an unprecedentedly large-scale, adventurist war but also to stand on the very brink of final collapse. Viewed in terms of historical development, Japan is no longer an ascending nation; war cannot achieve the prosperity sought by its ruling class but will instead bring about the opposite of its aspirations—the demise of Japanese imperialism itself. This is what is meant by the retrogressive character of Japan's war. Coupled with this retrogression, and further compounded by Japan's distinctive character as a militarily feudalistic imperialism, arises the peculiar barbarity of its war. Consequently, this war will maximally exacerbate class antagonisms within Japan, antagonism between the Japanese and Chinese nations, and antagonism between Japan and the majority of nations worldwide. The retrogressive and barbarous nature of Japan's war constitutes the principal basis for its inevitable defeat.

Moreover, third, although Japan's war is waged on the foundation of its considerable military, economic, and political-organizational strength, it is simultaneously waged on a foundation of inherent insufficiency. While Japan's military, economic, and political-organizational capacities are strong, these strengths are deficient in quantitative terms. Japan is a comparatively small country; its human resources, military manpower, financial reserves, and material supplies are all insufficient to sustain a prolonged war. Japan's rulers seek to resolve this difficulty through war, but they too shall attain the opposite of their aspirations: that is, by initiating war to solve this difficulty, they will instead intensify their difficulties through war, and the conflict will consume even what they originally possessed.

Finally, fourth, although Japan may receive assistance from international fascist states, it simultaneously cannot avoid encountering an international opposition force exceeding the magnitude of such assistance. This latter force will gradually grow, ultimately not only neutralizing the former's supportive capacity but also exerting pressure upon Japan itself. This is the law that "those who lack moral principle find scant support"—a law generated by the very nature of Japan's war.

In summary, Japan's strength lies in the power of its war-making capacity, while its weaknesses reside in the retrogressive and barbarous essence of its war, in its insufficiency of human and material resources, and in the unfavorable international situation of being isolated. Such are the characteristics of Japan.

(11) Regarding China:

First, we are a semi-colonial, semi-feudal country. From the Opium War [10], through the Taiping Heavenly Kingdom [11], the Hundred Days' Reform of 1898 [12], the Revolution of 1911 [13], up to the Northern Expedition, all revolutionary or reformist movements aimed at eliminating our semi-colonial, semi-feudal status suffered severe setbacks; consequently, this status remains unchanged. We remain a weak nation; in military strength, economic capacity, and political-organizational power, we appear inferior to the enemy on all fronts. The inevitability of war and China's inability to achieve swift victory also find their basis herein.

However, second, China's liberation movement over the past century has accumulated to the present day and now differs fundamentally from any prior historical period. Although various internal and external reactionary forces have inflicted severe setbacks upon the liberation movement, they have simultaneously tempered the Chinese people. Though China's military, economic, political, and cultural conditions today are not as strong as Japan's, when compared with China's own historical past, they now contain more progressive factors than at any previous period. The Communist Party of China and the armed forces under its leadership represent these progressive factors. China's present war of liberation derives its possibility of protracted struggle and ultimate victory precisely from this progressive foundation. China is a nation rising like the morning sun—a sharp contrast to the declining state of Japanese imperialism. China's war is progressive; from this progressiveness arises the just character of China's war. Because this war is just, it can mobilize national unity, evoke sympathy from the people of the enemy country, and win the support of the majority of nations worldwide.

Third, China is moreover a vast country—extensive in territory, rich in resources, populous, and possessing substantial military forces—capable of sustaining a prolonged war. This, too, stands in direct contrast to Japan.

Finally, fourth, the broad international support generated by the progressive and just character of China's war stands in exact opposition to Japan's isolation due to its lack of moral principle.

In summary, China's weakness lies in the relative inferiority of its war-making capacity, while its strengths reside in the progressive and just essence of its war, in its status as a large country capable of sustaining protracted conflict, and in the favorable international situation of receiving widespread support. Such are the characteristics of China.

(12) Thus we observe: Japan possesses superior military strength, economic capacity, and political-organizational power, yet its war is retrogressive and barbarous, its human and material resources are insufficient, and its international situation is unfavorable. China, by contrast, possesses comparatively weaker military, economic, and political-organizational capacities, yet it stands in a progressive historical era; its war is progressive and just; it possesses the condition of being a large country sufficient to sustain protracted war; and the majority of nations worldwide will support China.

These constitute the fundamental, mutually contradictory characteristics of the Sino-Japanese War. These characteristics determine—and continue to determine—all political policies and military strategies and tactics on both sides; they determine—and continue to determine—the protracted nature of the war and the fact that ultimate victory belongs to China, not to Japan. War is precisely the contest of these characteristics. In the course of the war, these characteristics will undergo transformations in accordance with their inherent nature, and all phenomena will emerge therefrom. These characteristics exist objectively; they are not fabrications intended to deceive. They constitute the complete, fundamental elements of the war—not fragmentary or incomplete excerpts. They permeate every issue, great or small, and every stage of operations on both sides; they are not optional or dispensable.

Any analysis of the Sino-Japanese War that overlooks these characteristics will inevitably err; even if certain views temporarily gain credence and appear plausible, the actual course of the war will ultimately prove them incorrect. We now proceed, on the basis of these characteristics, to address all the questions we intend to examine.

III. Refutation of the Theory of National Subjugation

(13) Proponents of the theory of national subjugation focus solely on the single factor of the relative strength of the enemy and ourselves. In the past, they declared: "If we resist, we shall surely perish"; now they assert: "If we continue to fight, we shall certainly perish." If we merely respond that, although the enemy is strong, it is a small country, and although China is weak, it is a large country, this will not suffice to refute them. They can cite historical evidence—such as the Mongol conquest of the Song dynasty or the Qing conquest of the Ming—to demonstrate that a small but strong nation can indeed subjugate a large but weak one, and moreover, that a backward nation can conquer a progressive one. If we reply that these are ancient examples and thus inapplicable, they can then invoke the British conquest of India to prove that a small but strong capitalist country can indeed subjugate a large but weak backward nation. Therefore, we must advance additional grounds to silence all proponents of the subjugation theory, to convince them intellectually, and to provide all engaged in propaganda work with sufficient arguments to persuade those who remain unclear or unsteady, thereby consolidating their confidence in the War of Resistance.

(14) What, then, are these additional grounds? They reside in the characteristics of the present epoch. The concrete manifestation of this characteristic is Japan's retrogression and isolation versus China's progress and widespread support.

(15) Our war is no ordinary conflict; it is a war waged between China and Japan in the 1930s. Regarding our enemy: first, it is a moribund imperialist power, already situated in an era of retrogression. It differs not only from Britain during its conquest of India—when Britain was still in the progressive phase of capitalism—but also from Japan itself during the First World War two decades prior. This war has been launched on the eve of the great collapse of world imperialism, particularly of the fascist states; the enemy has embarked upon this desperate, adventurist war precisely for this reason. Consequently, the inevitable outcome of the war will not be the subjugation of China but the downfall of Japan's imperialist ruling clique—an inescapable necessity. Furthermore, at the time Japan initiated this war, the nations of the world were either already engaged in conflict or on the verge of war; all were preparing to resist barbarous aggression. China, as a nation, shares vital interests with the majority of the world's countries and peoples—this constitutes the root cause of the opposition

that Japan has already provoked and will continue to deepen among the majority of nations and peoples worldwide.

(16) What of China? It can no longer be compared with any prior historical period. Its semi-colonial, semi-feudal character defines it as a weak nation. Yet simultaneously, it exists in a historically progressive era—this constitutes the principal basis for its capacity to defeat Japan. To say that the War of Resistance is progressive does not signify ordinary, general progress; nor does it refer to the kind of progress exhibited in Abyssinia's war against Italian aggression; nor is it equivalent to the progress of the Taiping Heavenly Kingdom or the Revolution of 1911. Rather, it denotes the progress of China today. Wherein lies China's present progress? It lies in the fact that China is no longer a wholly feudal society; it has developed capitalism, a bourgeoisie and a proletariat, a broad populace that has awakened or is awakening, a Communist Party, politically progressive armed forces—the Chinese Red Army under Communist leadership—and decades of accumulated revolutionary experience, particularly the seventeen years since the founding of the Communist Party of China. These experiences have educated the Chinese people and the Chinese political parties; today, they precisely constitute the foundation for united resistance against Japan. If, in Russia, the victory of 1917 would not have been possible without the experience of 1905, then we may likewise assert that without the experience accumulated over the past seventeen years, victory in the War of Resistance would likewise be impossible. These are the domestic conditions.

The international conditions, which ensure that China is not isolated in this war, represent something unprecedented in history. Whether in China's past wars or India's, all were fought in isolation. Only today does China encounter an unprecedentedly vast and profound popular movement worldwide, along with its support for China. The Russian Revolution of 1917 also received international assistance, enabling the Russian workers and peasants to triumph; yet the scale of that assistance was not as extensive as today's, nor was its nature as profound. Today, the global popular movement is developing on an unprecedented scale and with unprecedented depth. The existence of the Soviet Union constitutes an especially crucial factor in contemporary international politics; it will inevitably provide China with enthusiastic assistance—a phenomenon entirely absent two decades ago. All these factors have created, and continue to create, indispensable conditions for China's ultimate victory.

Although substantial, direct assistance has not yet materialized and must await the future, China's progressive character and status as a large nation enable it to prolong the war, thereby promoting and awaiting international support.

(17) Adding to this the condition that Japan is a small country—limited in territory, resources, population, and military forces—whereas China is a large country—vast in territory, abundant in resources, populous, and possessing substantial military forces—we observe that, beyond the contrast of strength versus weakness, there exists the contrast of small/retrogressive/isolated versus large/progressive/widely supported. This constitutes the basis for China's absolute immunity from subjugation. Although the disparity in strength dictates that Japan can, for a certain period and to a certain extent, run rampant in China, and that China must inevitably traverse a difficult path—making the War of Resistance a protracted rather than a swift decisive conflict—the contrast of small/retrogressive/isolated versus large/progressive/widely supported simultaneously dictates that Japan cannot prevail indefinitely; it will inevitably suffer ultimate defeat, while China will never perish and will assuredly achieve final victory.

(18) Why did Abyssinia perish? First, it was not merely weak but also small. Second, it was less progressive than China: it was an ancient society transitioning from slavery to serfdom, lacking capitalism, lacking a bourgeois political party, let alone a Communist Party; it possessed no military comparable to China's, much less forces akin to the Eighth Route Army. Third, it could not await international assistance; its war was fought in isolation. Fourth—and most critically—its leadership committed errors in conducting the war against Italy. For these reasons, Abyssinia perished. Nevertheless, Abyssinia still maintained considerable guerrilla warfare; had it persisted, it might have leveraged future global transformations to restore its homeland.

(19) If proponents of the subjugation theory invoke the history of failures in China's modern liberation movements to "prove" that "resistance will surely lead to subjugation" or "continued fighting will surely lead to subjugation," our reply remains: the epoch is different. China itself, Japan's internal situation, and the international environment are all unlike the past. Japan is stronger than before; China's semi-colonial, semi-feudal status remains unchanged, and its strength remains comparatively weak—

this is a grave circumstance. Japan can still temporarily control its domestic population and exploit international contradictions as instruments for its invasion of China—these are facts. However, in the course of a prolonged war, opposite transformations will inevitably occur. Though this is not yet fact, it will inevitably become fact in the future. This is precisely what the proponents of subjugation theory disregard.

What of China? Not only does it now possess new people, new political parties, new armed forces, and new policies of resistance—marking a profound difference from a decade or more ago—but all these elements will inevitably continue to advance. Although historical liberation movements have repeatedly suffered setbacks, preventing China from accumulating greater strength for today's War of Resistance—a profoundly regrettable historical lesson from which we must never again allow ourselves to undermine any revolutionary force—nevertheless, building upon the existing foundation and adding extensive effort, China can assuredly advance gradually and strengthen its capacity for resistance. The great anti-Japanese national united front constitutes the general direction of this effort.

Regarding international assistance: although substantial and direct aid is not yet visible, the international situation has fundamentally changed from the past; substantial and direct assistance is now in preparation. The failures of China's numerous modern liberation movements each had objective and subjective causes; none can be equated with today's circumstances. Although many difficult conditions exist today—dictating that the War of Resistance will be arduous, such as the enemy's strength, our weakness, the enemy's difficulties only just beginning, and our progress still insufficient—nevertheless, the favorable conditions for defeating the enemy are numerous. With the addition of subjective effort, these difficulties can be overcome and victory secured. These favorable conditions surpass any prior historical period; this is the reason why the War of Resistance will not share the same fate of failure as historical liberation movements.

IV. Compromise or Resistance? Corruption or Progress?

(20) The lack of basis for the theory of national subjugation has been demonstrated above. However, there are many others who are not proponents of subjugation theory; they are patriotic volunteers, yet they harbor profound anxieties regarding the current situation. Their concerns are twofold: first, fear of compromise with Japan; second, doubt that politics can progress. These two worrying issues are being discussed among broad masses of people without finding a基点 (basis for resolution). We shall now examine these two questions.

(21) As stated earlier, the problem of compromise has its social roots; so long as this social foundation exists, the issue of compromise will inevitably arise. However, compromise will not succeed. To prove this, we need only seek grounds from three aspects: Japan, China, and the international situation. First, regarding Japan: At the outset of the War of Resistance, we estimated that an opportunity for cultivating an atmosphere of compromise would arise—namely, after the enemy occupied North China and the Jiang-Zhe region, they might employ means of persuading surrender. Later, this tactic indeed appeared; but the crisis soon passed. One reason was that the enemy adopted a universal policy of barbarity, implementing open plunder. If China surrenders, everyone becomes a subjugated slave. The enemy's policy of plunder—that is, of annihilating China—is applied universally to the Chinese people, both materially and spiritually; not only to the lower masses, but also to the upper strata— though naturally, slightly more politely toward the latter, yet this is only a difference of degree, not of principle. Broadly speaking, the enemy is transplanting the old methods used in the three northeastern provinces into the interior. Materially, they plunder the clothing and food of ordinary people, causing the broad masses to cry out from hunger and cold; they plunder means of production, causing China's national industry to be destroyed and enslaved. Spiritually, they crush the national consciousness of the Chinese people. Under the Japanese flag, every Chinese can only be a compliant subject, a beast of burden; not a shred of Chinese spirit is permitted. The enemy's barbarous policy will be applied even deeper into the interior. Their appetite is voracious; they are unwilling to stop the war. The policy declared by the Japanese Cabinet on January 16, 1938 [14] is being resolutely implemented to this day, and cannot but be implemented; this has enraged the Chinese people of all strata. This arises from the retrogressive and barbarous nature of the enemy's war; "inescapable

calamity" [注] thus forms absolute hostility. Estimating that at some opportunity, the enemy's means of persuading surrender will appear again, certain proponents of subjugation theory will again stir蠕蠕 (like worms), and inevitably collude with certain international elements (there are such people within Britain, America, and France, especially among the British upper strata), acting as accomplices. However, the trend of the general situation is that surrender is impossible; the resoluteness and special barbarity of Japan's war dictate this aspect of the problem.

(22) Second, regarding China: There are three factors sustaining China's persistence in resistance: First, the Communist Party, which is a reliable force leading the people in resistance. Second, the Kuomintang, because it relies on Britain and America; if Britain and America do not order it to surrender, it will not surrender. Third, other political parties, the majority of which oppose compromise and support resistance. These three forces unite with one another; whoever seeks compromise stands on the side of the traitors, and everyone may execute them. All who are unwilling to be traitors cannot but unite to persist in resistance to the end; thus, compromise is practically difficult to succeed.

(23) Third, regarding the international aspect: Except for Japan's allies and certain elements within the upper strata of capitalist countries, the rest are unfavorable to China's compromise and favorable to China's resistance. This factor affects China's hopes. Today, the entire nation has a hope, believing that international forces will gradually increase assistance to China. This hope is not empty; especially the existence of the Soviet Union encourages China's resistance. The unprecedentedly powerful socialist Soviet Union has always shared weal and woe with China. The Soviet Union is fundamentally opposite to those profit-seeking elements among the upper strata of all capitalist countries; it takes assisting all weak nations and revolutionary wars as its mission. China's war's non-isolation is not only generally built upon entire international assistance but specifically built upon Soviet assistance. China and the Soviet Union are geographically close; this point aggravates Japan's crisis and facilitates China's resistance. The geographical proximity of China and Japan aggravates the difficulties of China's resistance. However, the geographical proximity of China and the Soviet Union is a favorable condition for China's resistance.

(24) From this, we may conclude: The crisis of compromise exists, but it can be overcome. Because even if the enemy's policy may undergo some degree of change, its fundamental change is impossible. Internally, China has social roots for compromise, but the majority oppose compromise. International forces also have a portion sponsoring compromise, but the main forces sponsor resistance. Combining these three factors enables overcoming the compromise crisis and persisting in resistance to the end.

(25) Now we address the second question. The improvement of domestic politics is inseparable from the persistence of resistance. The more politics improve, the more resistance can be persisted in; the more resistance is persisted in, the more politics will improve. But fundamentally, it depends on persisting in resistance. The various undesirable phenomena of the Kuomintang seriously exist; the historical accumulation of these unreasonable factors causes great anxiety and depression among broad patriotic volunteers. However, the experience of resistance has already proven that ten months of Chinese people's progress equals decades of progress in the past; there is no basis for pessimism. The corruption phenomena accumulated by history, although seriously hindering the speed of growth of the people's resistance power, reducing victory in war, and causing losses in war, nevertheless, the general situation of China, Japan, and the world does not permit the Chinese people not to progress. Due to the existence of factors hindering progress—namely, corruption phenomena—this progress is slow. Progress and the slowness of progress are two characteristics of the current situation; the latter characteristic is very incompatible with the urgent demands of war, which is where patriotic volunteers are greatly worried. However, we are in a revolutionary war; revolutionary war is an antitoxin; it will not only exclude the enemy's poisonous flames but also cleanse its own filth. All just revolutionary wars possess great power; they can transform many things or open the road for transforming things. The Sino-Japanese War will transform both China and Japan; so long as China persists in resistance and persists in the united front, it will certainly transform old Japan into new Japan, and old China into new China; the people and things of both China and Japan will be transformed during and after this war. It is proper for us to link resistance with national construction. To say Japan can also be transformed means that the侵略 war (aggressive war) of Japan's rulers will proceed to failure, possibly causing revolution among the Japanese people. The day of the Japanese people's revolutionary victory is the time of Japan's transformation. This is closely linked to China's resistance; this prospect should be seen.

V. The Theory of National Subjugation is Wrong, and So is the Theory of Quick Victory

(26) We have already conducted comparative research on several basic contradictory characteristics between enemy and ourselves—strength versus weakness, large versus small, progress versus retrogression, much assistance versus little assistance—refuting the theory of national subjugation and answering why compromise is not easy and why political progress is possible. Proponents of subjugation theory emphasized only the single contradiction of strength versus weakness, exaggerating it as the basis for the entire problem, while ignoring other contradictions. Their mentioning only the point of strength versus weakness is their one-sidedness; their exaggerating this one-sided thing into the whole is their subjectivity. Therefore, speaking generally, they are without basis and are erroneous. Regarding those who are not proponents of subjugation theory, nor consistent pessimists, but are merely confused for a time by the enemy-versus-us strength situation or domestic corruption phenomena at a certain moment or locality, and temporarily develop a pessimistic psychology— we must also point out to them that the source of their viewpoint is also a tendency toward one-sidedness and subjectivity. However, their correction is easier; a mere reminder will make them understand, because they are patriotic volunteers, and their error is temporary.

(27) However, proponents of the theory of quick victory are also incorrect. They either fundamentally forget the contradiction of strength versus weakness, remembering only other contradictions; or they exaggerate China's strengths away from the true situation, turning them into something else; or they substitute the strength-versus-weakness phenomenon of a certain moment or locality for the strength-versus-weakness phenomenon of the whole, letting one leaf obscure Mount Tai, yet considering themselves correct. In short, they lack the courage to admit the fact that the enemy is strong and we are weak. They often obliterate this point, thereby obliterating one aspect of truth. They also lack the courage to admit the limited nature of their own strengths, thereby obliterating another aspect of truth. Thus, they commit errors great or small; here too, subjectivity and one-sidedness are at fault. These friends have good hearts; they are also patriotic volunteers. But "the gentleman's ambition is indeed great" [注], the gentleman's view is incorrect; if acted upon, one will certainly hit a wall. Because estimation does not conform to truth, action cannot achieve its purpose; forcing it through will result in defeated armies and a perished nation, the outcome no different from defeatists. Therefore, this is also unacceptable.

(28) Do we deny the danger of national subjugation? We do not deny it. We admit that before China lie two possible futures—liberation and subjugation—and both are struggling fiercely. Our task is to realize liberation and avoid subjugation. The conditions for realizing liberation are fundamentally China's progress, simultaneously added to the enemy's difficulties and the world's assistance. We differ from proponents of subjugation theory; we objectively and comprehensively admit that both possibilities of subjugation and liberation exist simultaneously, emphasizing that the possibility of liberation holds advantage and the conditions for reaching liberation, and striving for these conditions. Proponents of subjugation theory subjectively and one-sidedly admit only one possibility of subjugation, denying the possibility of liberation, even less pointing out the conditions for liberation and striving for these conditions. We also admit the tendency toward compromise and corruption phenomena, but we also see other tendencies and other phenomena, and point out that the latter will gradually gain advantage over the former, and the two are struggling fiercely; and we point out the conditions for the latter's realization, striving to overcome the tendency toward compromise and transform corruption phenomena. Therefore, we are not pessimistic, whereas pessimistic people are the opposite.

(29) Nor do we dislike quick victory; who would not approve of driving the "Japanese devils" out tomorrow morning? But we point out that without certain conditions, quick victory exists only in the mind; objectively, it does not exist; it is merely fantasy and false reasoning. Therefore, we objectively and comprehensively estimate all enemy-versus-us situations, pointing out that only strategic protracted war is the sole path to striving for final victory, while rejecting the baseless theory of quick victory. We advocate striving for all conditions necessary for achieving final victory; the more conditions possessed, the sooner possessed, the greater the grasp of victory, the earlier the time of victory. We believe only thus can the process of war be shortened, while rejecting the theory of quick victory that seeks cheap gains and values empty talk.

V. Why a Protracted War?

(30) We now proceed to examine the question of protracted war. The question "Why a protracted war?" can only be correctly answered by basing ourselves on the totality of the fundamental factors comparing the enemy and ourselves. For example, to state merely that the enemy is a powerful imperialist country and we are a weak semi-colonial, semi-feudal country carries the danger of falling into the theory of national subjugation. For to oppose weakness with strength alone, whether in theory or in practice, cannot produce a protracted result. To consider size alone, or progress versus retrogression, or much assistance versus little assistance, is likewise insufficient. It is common for the large to be merged with the small, or the small with the large. A progressive country or thing, if its strength is not formidable, is often annihilated by a large but retrogressive country or thing. Much assistance versus little assistance is an important factor, but it is an accessory factor; its magnitude of effect depends upon the fundamental factors inherent in the enemy and ourselves. Therefore, when we say the War of Resistance Against Japan is a protracted war, this is a conclusion derived from the interrelation of all factors comparing the enemy and ourselves. The enemy is strong and we are weak; we face the danger of perishing. But the enemy has other shortcomings, and we have other advantages. The enemy's advantages can be weakened through our efforts, and his shortcomings can be expanded through our efforts. Conversely, our advantages can be strengthened through our efforts, and our shortcomings can be overcome through our efforts. Therefore, we can achieve ultimate victory and avoid perishing, while the enemy will ultimately be defeated and cannot avoid the collapse of his entire imperialist system.

(31) Since the enemy has only one advantage and all else are shortcomings, while we have only one shortcoming and all else are advantages, why does this not result in a balanced outcome, but instead creates the present situation of enemy advantage and our disadvantage? Clearly, this problem cannot be viewed formalistically. The fact is that the disparity in the degree of strength between enemy and ourselves is currently too great; the enemy's shortcomings have not yet developed, nor can they develop for the time being, to the extent necessary to neutralize the factors of his strength; our advantages have not yet developed, nor can they develop for the time being, to the extent necessary to supplement the

factors of our weakness. Therefore, balance cannot appear; what appears is imbalance.

(32) The enemy is strong and we are weak; the enemy has the advantage and we have the disadvantage. Although this situation undergoes some change through our efforts to persist in resistance and persist in the united front, a fundamental change has not yet occurred. Therefore, at a certain stage of the war, the enemy can achieve a certain degree of victory, while we will suffer a certain degree of defeat. However, why are both enemy and ourselves limited to a certain degree of victory or defeat within this certain stage, unable to exceed this and achieve total victory or total defeat? This is because, first, the original condition of enemy strength and our weakness is relative, not absolute; second, our efforts to persist in resistance and persist in the united front further create this relative situation. Speaking of the original condition: although the enemy is strong, his strength has already been diminished by other unfavorable factors, though not yet to the extent necessary to destroy his advantage; although we are weak, our weakness has already been supplemented by other favorable factors, though not yet to the extent necessary to change our disadvantage. Thus, it forms a situation where the enemy is relatively strong and we are relatively weak; the enemy has a relative advantage and we have a relative disadvantage. The strength, weakness, advantage, and disadvantage of both sides were originally not absolute; 加之 (added to this) the efforts of our persisting in resistance and persisting in the united front during the course of the war, the original situation of strength and weakness, advantage and disadvantage between enemy and ourselves changes even more. Consequently, enemy and ourselves are limited to a certain degree of victory or defeat within a certain stage, creating the situation of a protracted war.

(33) However, the situation continues to change. In the course of the war, so long as we can employ correct military and political strategies, commit no errors of principle, and exert the utmost effort, the enemy's unfavorable factors and our favorable factors will both develop with the prolongation of the war. They will inevitably continue to change the original degree of strength between enemy and ourselves, and continue to transform the situation of advantage and disadvantage between enemy and ourselves. When a new certain stage is reached, a great change in the degree of strength and the situation of advantage and disadvantage will occur, achieving the result of enemy defeat and our victory.

(34) At present, the enemy is still barely able to utilize his factors of strength; our resistance has not yet inflicted a fundamental weakening upon him. The factor of his insufficiency in manpower and material resources is not yet sufficient to halt his offensive; on the contrary, it is still sufficient to sustain his offensive to a certain degree. The factors sufficient to exacerbate class antagonism within his own country and national resistance within China—namely, the factor of the retrogressive and barbarous nature of the war—have not yet created a situation sufficient to fundamentally hinder his offensive. The factor of his international isolation is still in the process of change and development; it has not yet reached complete isolation. Many countries showing assistance to us still have arms capitalists and war material capitalists who, seeking profit alone, are supplying Japan with large quantities of war materials [15]; their governments [16] are also still unwilling to sanction Japan in practical methods together with the Soviet Union. All this dictates that our resistance cannot achieve quick victory, but can only be a protracted war. On the Chinese side, the factors of weakness manifested in military, economic, political, and cultural aspects, although having achieved some degree of progress during the ten months of resistance, are still far from the degree necessary to halt the enemy's offensive and prepare for our counteroffensive. Moreover, in terms of quantity, they have inevitably been somewhat diminished. Our various favorable factors, although all playing a positive role, still require tremendous effort to reach the degree sufficient to halt the enemy's offensive and prepare for our counteroffensive. Domestically, overcoming corruption phenomena and increasing the speed of progress; internationally, overcoming pro-Japan forces and increasing anti-Japan forces—these are not yet current realities. All this again dictates that the war cannot achieve quick victory, but can only be a protracted war.

VI. The Three Stages of Protracted War

(35) Since the Sino-Japanese War is a protracted war and ultimate victory will belong to China, it may be reasonably inferred that this protracted war will manifest concretely in three stages. The first stage is the period of the enemy's strategic offensive and our strategic defense. The second stage is the period of the enemy's strategic consolidation and our preparation for counteroffensive. The third stage is the period of our strategic counteroffensive and the enemy's strategic retreat. The specific conditions of the three stages cannot be predicted with certainty, but based on current conditions, certain general trends of the war can be indicated. The course of objective reality will be exceptionally rich and tortuously changing; no one can construct a "fortune-teller's almanac" of the Sino-Japanese War. However, outlining a contour of the war's trend is necessary for strategic guidance. Therefore, although what is outlined may not fully accord with future facts and will be corrected by facts, for the sake of conducting strategic guidance of the protracted war firmly and purposefully, the task of outlining a contour remains necessary.

(36) The first stage is not yet concluded. The enemy's intention is to occupy the three points of Guangzhou, Wuhan, and Lanzhou, and to link these three points. To achieve this objective, the enemy must deploy at least fifty divisions, approximately 1.5 million troops, over a period of one and a half to two years, at a cost exceeding 10 billion yen. Should the enemy penetrate so deeply, their difficulties will be extremely great, and the consequences will be unimaginable. As for attempting to fully occupy the Canton-Hankow Railway and the Sian-Lanchow Highway, this will entail an extremely hazardous war, and they may not fully achieve their intention. However, our operational plan should take as a basis the possibility that the enemy may occupy the three points and even certain areas beyond them, and may link them, thereby deploying for a protracted war; even if the enemy does so, we shall have means to cope. The form of warfare adopted by us in this stage is primarily mobile warfare, supplemented by guerrilla warfare and positional warfare. Although positional warfare was placed in a primary position during the first period of this stage due to subjective errors by the Kuomintang military authorities, viewed across the entire stage, it remains supplementary. In this stage, China has already formed a broad united front, achieving unprecedented unity. Although the enemy has employed and will continue to employ despicable and shameless means of persuading surrender,

attempting to realize their plan of quick decision and conquer China entirely without expending great effort, the past has already failed, and the future will hardly succeed. In this stage, although China has suffered considerable losses, there has simultaneously been considerable progress; this progress becomes the main foundation for continuing resistance in the second stage. In this stage, the Soviet Union has already provided substantial assistance to our country. On the enemy side, morale has begun to show signs of decline; the sharpness of the enemy army's offensive in the middle period of this stage is not as strong as in the initial period, and in the final period it will be even weaker than in the initial period. The enemy's finances and economy have begun to show signs of exhaustion; war-weariness among the people and soldiers has begun to occur; internal divisions within the war guidance group have begun to manifest "war weariness," growing pessimism regarding the war's prospects.

(37) The second stage may be termed the stage of strategic stalemate. At the end of the first stage, due to insufficient enemy manpower and our stubborn resistance, the enemy will be compelled to decide upon a limit for their strategic offensive; upon reaching this limit, they will halt their strategic offensive and shift to a stage of consolidating occupied territories. In this stage, the enemy's intention is to consolidate occupied territories, using deceptive methods of organizing puppet governments to claim them as their own, and to extract as much as possible from the Chinese people. However, before them they will encounter stubborn guerrilla warfare. Guerrilla warfare, taking advantage of the vacuum behind enemy lines in the first stage, will undergo universal development, establishing many base areas, fundamentally threatening the consolidation of enemy occupied territories; therefore, the second stage will still witness extensive warfare. In this stage, the form of our warfare is primarily guerrilla warfare, supplemented by mobile warfare. At this time, China will still be able to maintain large quantities of regular troops; however, on one hand, because the enemy adopts a strategic defensive posture in the occupied large cities and main roads, and on the other hand, because China's technical conditions cannot be perfected for the time being, it is difficult to launch a strategic counteroffensive swiftly. Apart from frontal defense forces, our army will shift largely behind enemy lines, deploying relatively dispersed, relying on all areas not occupied by the enemy, coordinating with armed masses, to wage extensive and fierce guerrilla warfare against enemy occupied territories, and as far as possible mobilize the enemy to destroy them in mobile warfare, as exemplified currently in Shanxi. The

warfare in this stage will be cruel; localities will encounter severe destruction. However, guerrilla warfare can be victorious; if done well, it is possible to cause the enemy to consolidate only about one-third of the occupied territories, while about two-thirds remain ours; this would be a great defeat for the enemy and a great victory for China. At that time, the entire enemy occupied territory will be divided into three types of areas: the first is the enemy's base areas, the second is the base areas of guerrilla warfare, and the third is guerrilla zones contested by both sides. The length of time of this stage depends on the degree of change in the increase or decrease of enemy and our strength and how the international situation changes; broadly speaking, we must prepare to pay a longer time, to endure this difficult journey. This will be a very painful period for China; economic difficulties and traitor sabotage will be two major problems. The enemy will vigorously engage in activities to rupture China's united front; all traitor organizations in enemy occupied territories will merge to form a so-called "unified government." Internally, due to the loss of large cities and the difficulties of war, wavering elements will loudly advocate the theory of compromise, and pessimistic sentiment will grow severely. At this time, our task lies in mobilizing the national masses, working heart and soul, unshakably persisting in war, expanding and consolidating the united front, excluding all pessimism and theory of compromise, advocating hard struggle, implementing new wartime policies, and enduring this difficult journey. In this stage, it is essential to call upon the whole nation to resolutely maintain a unified government, oppose splitting, systematically enhance operational techniques, transform the army, mobilize the entire people, and prepare for counteroffensive. In this stage, the international situation will change to become even more unfavorable for Japan; although there may be a shift in tone by Chamberlain [17]-type figures accommodating the so-called "accomplished fact" with "realism," the main international forces will shift toward further assistance to China. The Japanese threat to the South Seas and to Siberia will become more serious than in the past, possibly erupting into new wars. On the enemy side, the dozens of divisions trapped in the Chinese quagmire cannot be withdrawn. Extensive guerrilla warfare and the people's anti-Japanese movement will exhaust this large batch of Japanese troops, on one hand consuming them in large quantities, and on the other hand further increasing their homesickness, war-weariness, and even anti-war psychology, disintegrating this army spiritually. Although Japan's plunder in China cannot be said to be absolutely impossible to achieve anything, due to Japan's lack of capital and being trapped by guerrilla warfare, rapid and large-scale achievements are impossible. This second stage is the transition stage of the entire war, and will also be the most difficult period; however, it is the hub of transformation. Whether China becomes an independent country or becomes a colony is not decided by whether large cities are lost in the first stage, but by the degree of effort by the whole nation in the second stage. If we can persist in resistance, persist in the united front, and persist in protracted war, China will acquire the strength to transform weakness into strength in this stage. This is the second act of the three-act play of China's War of Resistance. Due to the efforts of all actors, the most wonderful finale can be performed well.

(38) The third stage is the stage of counteroffensive to recover lost territory. Recovering lost territory relies mainly on the strength that China itself prepared in the previous stage and continues to grow in this stage. However, one's own strength alone is not enough; it must also rely on assistance from international forces and changes within the enemy country; otherwise victory cannot be achieved, thus intensifying the task of China's international propaganda and diplomatic work. In this stage, war will no longer be strategic defense, but will become strategic counteroffensive; phenomenally, it will manifest as strategic offensive; it will no longer be strategic interior lines, but will gradually become strategic exterior lines. Only when fighting reaches the Yalu River will this war be considered concluded. The third stage is the final stage of the protracted war; the so-called persisting in war to the end means walking the full course of this stage. The main form of warfare adopted by us in this stage will still be mobile warfare, but positional warfare will be elevated to an important position. If positional defense in the first stage could not be regarded as important due to conditions at the time, then positional attack in the third stage, due to changed conditions and task requirements, will become quite important. Guerrilla warfare in this stage will still assist mobile warfare and positional warfare in its strategic coordinating role, differing from becoming the main form in the second stage.

(39) Thus it is seen that the long-term nature of the war and the ensuing cruelty are obvious. The enemy cannot swallow China entirely, but can occupy many places in China for a considerable length of time. China cannot swiftly expel Japan, but most of the land will still remain China's.

In the end, the enemy is defeated and we are victorious, but a difficult journey must be traversed.

(40) The Chinese people will receive excellent tempering in such a long and cruel war. The various political parties participating in the war will also be tempered and tested. The united front must be persisted in; only by persisting in the united front can war be persisted in; only by persisting in the united front and persisting in war can ultimate victory be had. If indeed this is so, all difficulties can be overcome. After crossing the difficult journey of war, the smooth path of victory will arrive; this is the natural logic of war.

(41) In the three stages, the change in enemy and our strength will advance along the following path. In the first stage, the enemy has the advantage and we have the disadvantage. This disadvantage of ours must be estimated to undergo two different changes from before the War of Resistance to the end of this stage. The first is downward. China's original disadvantage, through the consumption of the first stage, will become more serious; this is the reduction of land, population, economic strength, military strength, and cultural institutions. At the end of the first stage, this may be reduced to a considerable extent, especially in the economic aspect. This point will be utilized by people as a basis for the theory of national subjugation and the theory of compromise. However, it must be seen that there is a second change, namely upward change. This is experience in war, progress of the army, progress of politics, mobilization of the people, development of new directions in culture, emergence of guerrilla warfare, growth of international assistance, etc. In the first stage, downward things are old quantity and quality, mainly manifesting in quantity. Upward things are new quantity and quality, mainly manifesting in quality. This second change gives us the basis for being able to persist and achieve ultimate victory.

(42) In the first stage, there are also two changes on the enemy side. The first is downward, manifesting in: casualties of hundreds of thousands, consumption of weapons and ammunition, decline of morale, dissatisfaction among the domestic population, reduction of trade, expenditures of over 10 billion yen, blame from international public opinion, etc. This aspect again gives us the basis for being able to persist and achieve ultimate victory. However, the enemy's second change must also be estimated, namely upward change. That is, he has expanded territory, population, and resources. On this point, there arises again the basis for our War of Resistance being a protracted war and not a quick decisive war, and it will also be utilized by some people as a basis for the theory of national subjugation and the theory of compromise. However, we must estimate the temporary and partial nature of this upward change of the enemy. The enemy is an imperialist on the verge of collapse; the land he occupies in China is temporary. The fierce development of Chinese guerrilla warfare will actually limit his occupied zones to narrow areas. Moreover, the enemy's occupation of Chinese land has generated and deepened contradictions between Japan and foreign countries. Furthermore, based on the experience of the three northeastern provinces, Japan can generally only be in a period of capital expenditure for a considerable length of time, not a period of harvest. All these are again the basis for us to refute the theory of national subjugation and the theory of compromise and establish the theory of protracted war and the theory of ultimate victory.

VI. The Three Stages of Protracted War (Part 6B)

(43) In the second stage, the aforementioned changes on both sides will continue to develop. The specific circumstances cannot be predicted with certainty, but broadly speaking, Japan will continue downward while China continues upward [18]. For example, Japan's military strength and financial resources will be heavily consumed by China's guerrilla warfare; domestic discontent among the people will grow further, morale will decline even more, and internationally it will feel increasingly isolated. China, meanwhile, will achieve further progress in politics, military affairs, culture, and popular mobilization; guerrilla warfare will develop further; economically, relying on small-scale industry in the interior and vast agriculture, there will be some degree of new development; international assistance will gradually increase, improving significantly compared to the current situation. This second stage may perhaps endure for a considerable length of time. During this period, the comparison of strength between enemy and ourselves will undergo enormous opposite changes: China will gradually rise, while Japan will gradually decline. At that time, China will emerge from inferiority, while Japan will emerge from superiority; first arriving at a position of balance, then arriving at a position where superiority and inferiority are reversed. Subsequently, China will broadly complete preparations for strategic counteroffensive and proceed to the stage of implementing counteroffensive and expelling the enemy from the country. It should be repeatedly emphasized: the so-called transformation from inferiority to superiority and the completion of counteroffensive preparations include the growth of China's own strength, the increase of Japan's difficulties, and the growth of international assistance. Combining these forces will form China's superiority and complete preparations for counteroffensive.

(44) Based on the state of political and economic imbalance in China, the strategic counteroffensive in the third stage will not, in its earlier period, assume a uniform posture nationwide, but rather a regional posture of rising and falling. The enemy's attempt to use various divisive means to rupture China's united front will not diminish in this stage; therefore, the task of internal unity in China becomes even more critical, ensuring that internal discord does not cause the strategic counteroffensive to be abandoned halfway. During this period, the international situation will change to become greatly favorable to China. China's task lies in utilizing this international situation to achieve its own thorough liberation and establish an independent democratic country, simultaneously assisting the world's anti-fascist movement.

(45) China's progression from inferiority to balance to superiority; Japan's progression from superiority to balance to inferiority; China's progression from defense to stalemate to counteroffensive; Japan's progression from offensive to consolidation to retreat—such is the process of the Sino-Japanese War, the inevitable trend of the Sino-Japanese War.

(46) Thus the question and conclusion are: Will China perish? Answer: It will not perish; ultimate victory belongs to China. Can China achieve swift victory? Answer: It cannot achieve swift victory; it must be a protracted war. Is this conclusion correct? I believe it is correct.

(47) Upon hearing this, proponents of the theory of national subjugation and the theory of compromise will again emerge to say: For China to progress from inferiority to balance, it requires military and economic strength equal to Japan's; to progress from balance to superiority, it requires military and economic strength exceeding Japan's; however, this is impossible; therefore, the aforementioned conclusion is incorrect.

(48) This is the so-called "theory of weapons determinism," a mechanistic tendency in the problem of war, an opinion that views problems subjectively and one-sidedly. Our opinion is opposite to this: we see not only weapons but also human strength. Weapons are an important factor in war, but not the decisive factor; the decisive factor is man, not things. The comparison of strength is not only a comparison of military and economic power, but also a comparison of human strength and morale. Military and economic power must be mastered by men. If the majority of Chinese people, the majority of Japanese people, and the majority of people in all countries of the world stand on the side of the War of Resistance Against Japan, then can the military and economic power forcibly mastered by a minority of Japanese people still be considered an advantage? If it is not an advantage, then does not China, mastering relatively inferior military and economic power, become superior? Without doubt, so long as China persists in resistance and persists in the united front, its military and economic power can gradually be strengthened. Meanwhile, our enemy, through prolonged war and the weakening of internal and external contradictions, will inevitably undergo opposite changes in its military and economic power. Under these circumstances, can China not become superior? Moreover, at present we cannot openly count the military and economic power of other countries as our own strength in large quantities; but in the future, can we not? If Japan has more enemies than China alone, if in the future one or several countries openly defend against or attack Japan with their considerable military and economic power, openly aiding us, then will not superiority lie even more on our side? Japan is a small country; its war is retrogressive and barbarous; its international position will become increasingly isolated. China is a large country; its war is progressive and just; its international position will become increasingly supported. All these factors, through long-term development, can they not definitively change the situation of superiority and inferiority between enemy and ourselves?

(49) Proponents of the theory of quick victory, meanwhile, do not understand that war is a contest of strength; before a certain change has occurred in the comparison of strength between the two sides of the war, they wish to conduct strategic decisive engagements, wishing to arrive at the path of liberation prematurely; this is also without basis. Implementing their opinion will inevitably lead to hitting a wall. Or it is merely empty talk for satisfaction, without preparation to truly act. Finally, Mr. Fact will emerge to splash a ladle of cold water on these empty talkers, proving them to be merely empty talkists who are greedy for cheap gains and wish to expend little effort for much harvest. This kind of empty talkism has existed in the past and exists now, but it is not yet numerous; when the war develops to the stalemate stage and counteroffensive stage, empty talkism may increase. However, simultaneously, if China suffers relatively large losses in the first stage, and the second stage drags on for a very long time, the theory of national subjugation and the theory of compromise will become even more widely prevalent. Therefore, our firepower should be directed primarily at the theory of national subjugation and the theory of compromise, while using secondary firepower to oppose the empty talkist theory of quick victory.

(50) The protracted nature of the war is determined, but exactly how many years the war will pass through, no one can predict with certainty; this depends entirely on the degree of change in the strength of enemy and ourselves. All who wish to shorten the war time have only one method: strive to increase own strength and reduce enemy strength. Specifically, strive to win more battles in operations, consume enemy forces, strive to develop guerrilla warfare, restrict enemy occupied territory to the smallest scope, strive to consolidate and expand the united front, unite national

strength, strive to build new armies and develop new military industry, strive to promote progress in politics, economy, and culture, strive for mobilization of people from all circles of workers, peasants, merchants, and students, strive to disintegrate enemy forces and win over enemy soldiers, strive for international propaganda to win international assistance, strive to win the assistance of the Japanese people and other oppressed nations. Only by doing all these things can the war time be shortened; beyond this, there cannot be any clever shortcuts or convenient methods.

VII. Interlaced Warfare, Permanent Peace, and Human Agency

A. The Interlaced Character of the War (§§51–56)

(51) We may affirm with certainty that the protracted War of Resistance Against Japan will constitute a glorious and exceptional chapter in the history of human warfare. One particularly distinctive feature is its *interlaced character* (*dog-toothed interlocking*), which arises from the contradictory factors of Japan's barbarity and insufficient manpower versus China's progress and vast territory. Interlaced warfare has occurred in history—for example, during the three-year civil war following the October Revolution in Russia. However, its distinctive feature in China lies in its exceptional duration and scale; this will surpass all historical precedents. This interlaced form manifests in the following several respects.

(52) *Interior Lines and Exterior Lines* — Strategically, the War of Resistance as a whole occupies a position of interior-line operations; however, the relationship between the main forces and guerrilla units is such that the main forces operate on interior lines while guerrilla units operate on exterior lines, creating a remarkable spectacle of pincer attacks against the enemy. The same applies to the relationships among various guerrilla zones: each guerrilla zone regards itself as operating on interior lines while treating other zones as exterior lines, thereby forming numerous additional pincer lines against the enemy. In the first stage of the war, the regular forces conducting strategic interior-line operations are retreating, whereas the guerrilla forces conducting strategic exterior-line operations will advance in great strides extensively into the enemy's rear; in the second stage, this advance will become even more vigorous, creating a singular pattern of simultaneous retreat and advance.

(53) *With Rear Areas and Without Rear Areas* — The main forces utilize the nation's general rear area while extending their operational lines to the outermost limits of enemy-occupied territory. Guerrilla forces, by contrast, detach themselves from the general rear area and extend their operational lines into the enemy's rear. However, within each guerrilla base area, there still exists its own small-scale rear area, upon which non-fixed operational lines are established. Distinct from this are the guerrilla units dispatched from each base area for temporary operations behind enemy lines in that zone: they possess neither a rear area nor fixed operational lines. *"Operations without a rear area"* is a characteristic of revolutionary

warfare in the new era, under conditions of vast territory, a progressive populace, and the presence of an advanced political party and advanced armed forces; it is not to be feared but rather to be actively promoted, as it yields great advantages.

(54) *Encirclement and Counter-Encirclement* — Viewed from the perspective of the war as a whole, owing to the enemy's strategic offensive and exterior-line operations, we occupy a position of strategic defense and interior-line operations; unquestionably, we find ourselves within the enemy's strategic encirclement. This constitutes the enemy's *first form of encirclement* against us. However, by employing numerically superior forces against enemy columns advancing toward us from strategic exterior lines along multiple routes, and by adopting the operational principle of exterior-line warfare at the campaign and tactical levels, we can place one or several of these advancing enemy columns within our own encirclement. This constitutes our *first form of counter-encirclement* against the enemy.

Furthermore, viewed from the perspective of individual guerrilla base areas behind enemy lines, each isolated base area finds itself surrounded on three or four sides by the enemy—the former exemplified by the Wutai Mountains, the latter by northwestern Shanxi. This constitutes the enemy's *second form of encirclement* against us. However, if we view the various guerrilla base areas in conjunction with one another, and further link them with the positions of the regular forces, we in turn encircle many enemy forces—for example, in Shanxi, we have already encircled the Tongpu Railway on three sides (east and west flanks and southern terminus) and encircled the city of Taiyuan on four sides; similar encirclements exist in many locations in Hebei, Shandong, and other provinces. This constitutes our *second form of counter-encirclement* against the enemy.

Thus, both sides employ two forms of encirclement against one another; broadly speaking, the situation resembles a game of Go (*weiqi*). The enemy's and our own campaign- and battle-level operations resemble capturing stones; the enemy's strongpoints (e.g., Taiyuan) and our guerrilla base areas (e.g., Wutai Mountains) resemble making "eyes" to secure living space. If we further include the global dimension of Go, there exists a *third form of mutual encirclement*: the relationship between the axis of aggression and the front of peace. The enemy employs the former to encircle China, the Soviet Union, France, Czechoslovakia, and other nations; we employ the latter to counter-encircle Germany, Japan, and Italy. However, our encirclement resembles the palm of the Buddha: it will transform into a Five-Elements Mountain spanning the cosmos, finally pressing down upon these modern-day Sun Wukongs—the fascist aggressors—so that they may never rise again [19].

If, through diplomacy, we can establish a Pacific Anti-Japanese Front, treating China as one strategic unit, the Soviet Union and other possible nations as additional strategic units, and the Japanese people's movement as yet another strategic unit, thereby forming an inescapable heavenly net that leaves the fascist Sun Wukong nowhere to flee—then the moment of the enemy's demise will have arrived. In reality, the day of Japanese imperialism's complete overthrow must coincide with the approximate completion of this heavenly net. This is by no means a jest, but rather the inevitable trend of the war.

(55) *Large Areas and Small Areas* — One possibility is that enemy-occupied regions will comprise the greater part of China proper, while intact Chinese-controlled regions will comprise the lesser part. This is one scenario. However, within the greater part occupied by the enemy—excluding the three northeastern provinces and similar areas—they can in practice occupy only major cities, main transportation arteries, and certain flatlands; judged by importance, these are first-class, but judged by area and population, they may constitute only the lesser part of enemy-occupied territory, whereas the extensively developed guerrilla zones conversely comprise the greater part. This is another scenario.

If we extend our scope beyond China proper to include Mongolia, Xinjiang, Qinghai, and Tibet, then in terms of area, the regions not lost by China still constitute the greater part, while enemy-occupied regions—including the three northeastern provinces—constitute only the lesser part. This is yet another scenario. Intact regions are of course important and should be vigorously developed—not only in political, military, and economic respects, but also culturally. The enemy has transformed our former cultural centers into culturally backward regions; we, in turn, must transform formerly backward cultural regions into new cultural centers. Simultaneously, the development of the vast guerrilla zones behind enemy lines is also critically important; all aspects of these zones should be developed, including cultural work.

Viewed comprehensively, China will see large rural areas transformed into progressive and enlightened regions, while small enemy-occupied zones—especially major cities—will temporarily become backward and dark regions.

(56) Thus it is evident that the protracted and extensive War of Resistance Against Japan is a war of interlaced character across military, political, economic, and cultural dimensions—a marvel in the history of warfare, a heroic undertaking of the Chinese nation, and a world-shaking enterprise. This war will not only influence China and Japan, greatly advancing the progress of both nations, but will also influence the world, promoting the progress of other nations—first and foremost, oppressed nations such as India. All Chinese people should consciously immerse themselves in this interlaced warfare; this is the form of warfare through which the Chinese nation seeks its own liberation, the distinctive form of a liberation war waged by a large semi-colonial nation in the 1930s and 1940s.

B. Fighting for Permanent Peace (§§57–58)

(57) The protracted nature of China's War of Resistance Against Japan is inseparable from the struggle for permanent peace in China and the world. No historical period has ever witnessed warfare as closely approximating permanent peace as the present. Since the emergence of social classes, human life over thousands of years has been filled with warfare; every nation has fought countless battles, either within ethnic groupings or between them. With the advent of capitalist society and the imperialist stage, warfare has become exceptionally extensive and exceptionally cruel. The First World War two decades ago was unprecedented in prior history, yet it was not the final war. Only the war that has now begun approximates a final war—that is, it approximates humanity's permanent peace.

Currently, one-third of the world's population is already engaged in warfare: consider Italy, then Japan; Abyssinia, then Spain; then China. The nations participating in warfare collectively comprise nearly six hundred million people—almost one-third of the world's total population. The distinctive features of the present war are its uninterrupted character and its approximation to permanent peace.

Why uninterrupted? After Italy fought Abyssinia, Italy then fought Spain, with Germany taking a share; then Japan fought China. Who will be next? Unquestionably, it will be Hitler fighting the major powers. *"Fascism is war"* [20]—this is absolutely correct. The current war will proceed without interruption until it develops into a world war; humanity's calamity of warfare is unavoidable.

Why, then, do we also say that this war approximates permanent peace? This war has erupted on the foundation of the general crisis of world capitalism initiated by the First World War; owing to this general crisis, capitalist nations are compelled into new warfare, and fascist states in particular are driven to undertake the adventure of new war. We can foresee that the outcome of this war will not be the salvation of capitalism, but rather its march toward collapse. This war will be larger and more cruel than the war two decades ago; all nations will inevitably be drawn into it; the war will be prolonged; humanity will suffer greatly.

However, owing to the existence of the Soviet Union and the heightened political consciousness of the world's peoples, this war will unquestionably give rise to great revolutionary wars to oppose all counter-revolutionary wars, thereby imparting to this war the character of a struggle for permanent peace. Even if a subsequent period of warfare remains, it will nevertheless be not far from the world's permanent peace. Once humanity eliminates capitalism, it will reach the era of permanent peace; thereafter, warfare will no longer be necessary. At that time, there will be no need for armies, warships, military aircraft, or poison gas. Henceforth, humanity will not witness warfare for hundreds of millions of years.

The revolutionary war that has already begun constitutes a component of this war for permanent peace. The war between China and Japan, comprising over five hundred million people, will occupy an important position within this war; the liberation of the Chinese nation will be achieved through this war. The liberated new China of the future is inseparable from the liberated new world of the future. Therefore, our War of Resistance Against Japan embodies the character of a struggle for permanent peace.

(58) Wars in history fall into two categories: one is just; the other is unjust. All progressive wars are just; all wars that impede progress are unjust. We Communists oppose all unjust wars that impede progress, but we do not

oppose progressive, just wars. With respect to the latter category of wars, we Communists not only do not oppose them but actively participate in them.

The former category of wars—for example, the First World War, in which both sides fought for imperialist interests—was resolutely opposed by Communists worldwide. The method of opposition was: prior to the outbreak of war, to strive with all might to prevent its outbreak; once war had erupted, to oppose war with war, to oppose unjust war with just war, whenever conditions permitted.

Japan's war is an unjust war that impedes progress; all peoples of the world, including the Japanese people, should oppose it and are indeed opposing it. China, by contrast—from the people to the government, from the Communist Party to the Kuomintang—has unanimously raised the banner of righteousness and is waging a national revolutionary war of resistance against aggression. Our war is sacred, just, progressive, and seeks peace. It seeks not merely the peace of one nation, but the peace of the world; not merely temporary peace, but permanent peace.

To achieve this purpose, we must fight to the death; we must be prepared to make every sacrifice; we must persist to the end and never cease until our objective is attained. Though the sacrifices are great and the duration long, the new world of permanent peace and permanent brightness already stands vividly before us. Our conviction in waging this war is founded upon this struggle for a new China and a new world of permanent peace and permanent brightness. Fascism and imperialism seek to prolong warfare indefinitely; we seek to bring warfare to an end in the not-too-distant future. To achieve this purpose, the majority of humanity must exert tremendous effort. The four hundred and fifty million Chinese constitute one-fourth of all humanity; if they can unite in effort, overthrow Japanese imperialism, and create a free and equal new China, their contribution to the struggle for permanent peace worldwide will unquestionably be immensely great.

This hope is not empty; the trajectory of global socio-economic development has already approached this point; it requires only the effort of the majority, and within several decades, the objective can certainly be achieved.

C. The Role of Conscious Activity in War (§§59–62)

(59) The foregoing discussion has explained *why* the war is protracted and *why* ultimate victory belongs to China—broadly speaking, addressing the questions of *"what is"* and *"what is not."* We now turn to examining the questions of *"how to do"* and *"how not to do."* How is protracted war to be conducted? How is ultimate victory to be secured? These are the questions we shall now address. To this end, we shall successively examine the following topics: the role of conscious activity in war; war and politics; political mobilization for the War of Resistance; the purpose of war; offense within defense; speed within protraction; exterior lines within interior lines; initiative; flexibility; planning; mobile warfare; guerrilla warfare; positional warfare; warfare of annihilation; warfare of attrition; the possibility of exploiting enemy vulnerabilities; the question of decisive engagements in the War of Resistance; and the soldiery and the people as the foundation of victory. We begin with the question of conscious activity.

(60) We oppose viewing problems subjectively—meaning that when a person's thinking is not grounded in or does not conform to objective facts, it is mere fantasy, false reasoning; if acted upon, it will lead to failure, and therefore must be opposed. However, all things must be done by people; protracted war and ultimate victory will not materialize without human action. To act, one must first, based on objective facts, derive ideas, principles, and opinions; propose plans, guidelines, policies, strategies, and tactics; only then can action be performed effectively. Ideas and the like are subjective; doing or acting is the subjective manifested in the objective—both constitute humanity's distinctive *conscious activity*. We term this conscious activity *"conscious initiative"*; it is the characteristic that distinguishes humans from things.

All thinking that is grounded in and conforms to objective facts is correct thinking; all doing or action based on correct thinking is correct action. We must promote such thinking and action; we must promote this kind of conscious initiative. The War of Resistance Against Japan aims to expel imperialism and transform old China into new China; this objective can be achieved only by mobilizing the entire Chinese people and fully developing their conscious initiative in resisting Japan. Sitting idle leads only to destruction; there can be no protracted war, nor ultimate victory.

(61) Conscious initiative is a distinctive characteristic of humanity. Humans manifest this characteristic especially intensely in warfare. The outcome of war is certainly determined by conditions on both sides—military, political, economic, geographical, the nature of the war, international assistance, and so forth—yet it is not determined *solely* by these conditions. These conditions alone merely create the *possibility* of victory or defeat; they do not, in themselves, determine the outcome. To determine the outcome, subjective effort must be added—this is the guidance and conduct of war, which constitutes *conscious initiative in warfare*.

(62) Those who guide warfare cannot seek victory beyond the limits permitted by objective conditions; however, they can and must, *within* the limits of objective conditions, dynamically strive for victory. The stage upon which war commanders operate must be built upon the permissions of objective conditions; yet, relying upon this stage, they can direct many vivid, magnificent, and heroic dramas.

Upon the given foundation of objective material conditions, commanders in the War of Resistance Against Japan must exert their power, lead the entire army, defeat those national enemies, transform the condition of our society and state—invaded and oppressed—into a free and equal new China. Herein lies the necessity and utility of our capacity for subjective guidance.

We do not endorse any commander in the War of Resistance who, disregarding objective conditions, becomes a reckless adventurer who charges about blindly; however, we must encourage every commander to become a brave and wise general. They must possess not only the courage to overwhelm the enemy, but also the capacity to master the changes and development of the entire war. Commanders swim in the vast ocean of war; they must not allow themselves to drown, but must decisively and methodically reach the opposite shore. Strategic and tactical principles as laws for guiding war are precisely the art of swimming in this ocean of war.

VIII. War and Politics; The Purpose of War; Operational Principles

A. War and Politics (§§63–65)

(63) "War is the continuation of politics." In this respect, war is politics; war itself is an action of a political nature. There has never been a war without political character since ancient times. The War of Resistance Against Japan is a revolutionary war of the entire nation; its victory cannot be separated from the political purpose of the war—expelling Japanese imperialism and establishing a free and equal new China; it cannot be separated from the general policy of persisting in resistance and persisting in the united front; it cannot be separated from the mobilization of the entire people; it cannot be separated from political principles such as unity between officers and soldiers, unity between the army and the people, and disintegrating the enemy forces; it cannot be separated from the effective implementation of united front policy; it cannot be separated from cultural mobilization; it cannot be separated from efforts to win international forces and the assistance of the people in the enemy country. In a word, war cannot be separated from politics for a single moment. If there are military personnel in the War of Resistance who tend to belittle politics, isolating war and becoming absolutists about war, that is erroneous and must be corrected.

(64) However, war has its particularity; in this respect, war is not equivalent to general politics. "War is the continuation of politics by special means." When politics develops to a certain stage and can no longer proceed in the old way, war breaks out to sweep away obstacles on the political road. For example, China's semi-independent status is an obstacle to the political development of Japanese imperialism; Japan wishes to sweep it away, so it launches a war of aggression. China? Imperialist oppression has long been an obstacle to China's bourgeois-democratic revolution, hence there have been many wars of liberation attempting to sweep away this obstacle. Japan now uses war to oppress, wishing to completely cut off the path of China's revolution, so China is compelled to wage the War of Resistance, determined to sweep away this obstacle. Once the obstacle is removed and the political purpose achieved, the war ends. If the obstacle is not swept away cleanly, the war must continue to carry it through. For example, if the task of resistance against Japan is unfinished and some wish to seek compromise, it will certainly not succeed; because even if compromise is reached for some reason, war will still arise, the broad masses will certainly not submit, and war must continue to carry through the political purpose of the war. Therefore, it can be said: *Politics is war without bloodshed; war is politics with bloodshed.*

(65) Based on the particularity of war, there is a special organization for war, a special method, and a special process. This organization is the army and everything attached to it. This method is the strategy and tactics for guiding war. This process is a special form of social activity in which opposing armies use strategy and tactics favorable to themselves and unfavorable to the enemy to engage in attack or defense. Therefore, the experience of war is special. All who participate in war must break away from ordinary habits and become accustomed to war before they can strive for victory in war.

B. Political Mobilization for the War of Resistance (§§66–67)

(66) Such a great national revolutionary war cannot be victorious without universal and deep political mobilization. Before the War of Resistance, there was no political mobilization for resistance; this was a great deficiency in China, and we had already lost a move to the enemy. After the War of Resistance began, political mobilization was also very far from universal, let alone deep. The majority of the people heard the news from the enemy's artillery fire and airplane bombs. This is also a kind of mobilization, but it was done by the enemy for us, not by ourselves. People in remote areas who cannot hear the sound of guns are still living quietly there. This situation must be changed; otherwise, the life-and-death war cannot be victorious. We must absolutely not lose another move to the enemy; on the contrary, we must greatly develop this move to defeat the enemy. This move is of absolute importance; inferiority in weapons and so forth is secondary; this move is truly of primary importance. Mobilizing the entire nation's common people creates a vast ocean in which the enemy will be drowned; it creates remedial conditions to compensate for defects in weapons and so forth; it creates the premise for overcoming all difficulties in war. To achieve victory, we must persist in resistance, persist in the united front, and persist in protracted war. However, all these cannot be separated from mobilizing the common people. To seek victory while neglecting political mobilization is called "heading south while wanting to go north" [36]; the result will inevitably be the cancellation of victory.

(67) What is political mobilization? First, it is to inform the army and the people of the political purpose of the war. Every soldier and every person must be made to understand why we are fighting and what fighting has to do with them. The political purpose of the War of Resistance is "to expel Japanese imperialism and establish a free and equal new China"; this purpose must be told to all military personnel and civilians alike before an upsurge of resistance can be created, enabling hundreds of millions of people to unite as one and contribute everything to the war. Second, merely explaining the purpose is not enough; we must also explain the steps and policies to achieve this purpose—that is, there must be a political program. Now we have the *Ten-Point Program for Resistance Against Japan and National Salvation* [22], and we also have the *Program for Resistance Against Japan and National Reconstruction* [23]; these should be popularized among the army and the people, and all army and people should be mobilized to implement them. Without a clear and specific political program, it is impossible to mobilize the entire army and entire people to resist Japan to the end. Third, how to mobilize? Rely on speech, rely on leaflets and proclamations, rely on newspapers and books, rely on drama and film, rely on schools, rely on mass organizations, rely on cadre personnel. Now there are some in the Kuomintang-controlled areas, but they are a drop in the ocean; moreover, the methods do not suit the taste of the masses, and the attitude is alienated from the masses; this must be 切实 (earnestly) changed. Fourth, one mobilization is not enough; political mobilization for the War of Resistance is constant. It is not reciting the political program to the common people; no one listens to such recitation; it must be linked to the development of the war situation, linked to the lives of soldiers and common people, turning political mobilization for the war into a constant movement. This is an absolutely great matter; war depends on this first and foremost for victory.

C. The Purpose of War (§§68–71)

(68) Here we are not speaking of the political purpose of war; the political purpose of the War of Resistance is "to expel Japanese imperialism and establish a free and equal new China," which was stated earlier. Here we are speaking of what is called war as politics with bloodshed, the war of two armies killing each other—what is its fundamental purpose? The purpose of war is nothing else but *"to preserve oneself and eliminate the enemy"* (eliminating the enemy means disarming the enemy, that is, so-called "depriving the enemy of resistance," not completely eliminating their physical bodies). In ancient war, spears and shields were used: the spear is for offense, to eliminate the enemy; the shield is for defense, to preserve oneself. Until today's weapons, this is still the continuation of the two. Bombers, machine guns, long-range artillery, poison gas are developments of the spear; air raid shelters, steel helmets, concrete works, gas masks are developments of the shield. The tank is a new weapon combining the two into one. Offense is the main means of eliminating the enemy, but defense cannot be discarded either. Offense is directly for eliminating the enemy, but also for preserving oneself, because if the enemy is not eliminated, oneself will be eliminated. Defense is directly for preserving oneself, but it is also a means to assist offense or prepare to shift to offense. Retreat belongs to the category of defense; it is the continuation of defense; while pursuit is the continuation of offense. It should be pointed out: *in the purpose of war, eliminating the enemy is primary, preserving oneself is secondary*, because only by eliminating the enemy in large quantities can oneself be effectively preserved. Therefore, offense as the main means of eliminating the enemy is primary, while defense as an auxiliary means of eliminating the enemy and as a means of preserving oneself is secondary. In actual war, although there are many times when defense is primary and at other times offense is primary, yet viewing the war as a whole, offense is still primary.

(69) How to explain advocating brave sacrifice in war? Is this not contradictory to "preserving oneself"? It is not contradictory; it is opposite yet complementary. War is politics with bloodshed; it requires paying a price, sometimes a very great price. Partial and temporary sacrifice (not preserving), for the sake of the whole and permanent preservation. We say that offensive means, which are basically for eliminating the enemy, also contain the function of preserving oneself; the reason lies here. Defense must simultaneously have offense and should not be pure defense; this is also the reason.

(70) This purpose of war—preserving oneself and eliminating the enemy—is the essence of war; it is the basis of all war actions, from technical actions to strategic actions, all implement this essence. The purpose of war is the basic principle of war; all technical, tactical, campaign, and strategic principles and principles cannot be separated from it even a little. What is the meaning of the shooting principle of "conceal body, develop firepower"? The former is for preserving oneself, the latter is for eliminating the enemy. Because of the former, utilizing terrain and objects, adopting leapfrog movement, open order formations, and various methods arise. Because of the latter, clearing the field of fire, organizing fire nets, and various methods arise. Tactical assault teams, containment teams, reserve teams: the first is for eliminating the enemy, the second is for preserving oneself, the third is prepared to be used for two purposes according to the situation—either to reinforce the assault team or as a pursuit team, all are for eliminating the enemy; or to reinforce the containment team or as a covering team, all are for preserving oneself. In this way, all technical, tactical, campaign, and strategic principles, all technical, tactical, campaign, and strategic actions, cannot be separated from the purpose of war even a little; it permeates the whole of war and runs through the beginning and end of war.

(71) Guides at all levels of the War of Resistance cannot guide war by leaving aside the various mutually opposing basic factors between China and Japan, nor can they guide war by leaving aside this purpose of war. The various mutually opposing basic factors between the two countries unfold in the actions of war, becoming a struggle where each side strives to preserve oneself and eliminate the enemy. Our war lies in striving to win every battle, regardless of size; in striving to disarm a part of the enemy and damage a part of the enemy's personnel and material in every battle. Accumulating these achievements of partially eliminating the enemy becomes a great strategic victory, achieving the political purpose of finally driving the enemy out of the country, defending the motherland, and building a new China.

D. Offense Within Defense, Speed Within Protraction, Exterior Lines Within Interior Lines (§§72–77)

(72) Now we proceed to study the specific strategic policy in the War of Resistance. We have already said that the strategic policy of resistance against Japan is protracted war; yes, this is completely correct. But this is a general policy, not yet a specific policy. How to conduct protracted war concretely? This is the question we are now discussing. Our answer is: *In the first and second stages, namely the enemy's offensive and conservative stages, it should be offensive war in campaigns and battles within strategic defense, quick-decision war in campaigns and battles within strategic protraction, and exterior-line operations in campaigns and battles within* *strategic interior lines. In the third stage, it should be strategic counteroffensive war.*

(73) Because Japan is an imperialist strong country and we are a semi-colonial, semi-feudal weak country, Japan adopts the strategic policy of offensive, while we occupy the strategic position of defense. Japan attempts to adopt strategic quick-decision war; we should consciously adopt strategic protracted war. Japan uses several dozen divisions of its army with considerable combat strength (currently up to thirty divisions) and part of its navy to surround and blockade China from land and sea, and also uses air force to bomb China. Currently, the Japanese army has occupied the long front from Baotou to Hangzhou; the navy has reached Fujian and Guangdong, forming large-scale exterior-line operations. We occupy the position of interior-line operations. All these are caused by the characteristic of enemy strength and our weakness. This is one aspect of the situation.

(74) However, on the other hand, the opposite is true. Although Japan is strong, its military strength is insufficient. Although China is weak, its territory is vast, its population is large, and its soldiers are numerous. Here two important results arise. First, the enemy, with few soldiers confronting a large country, can only occupy a part of the big cities, main roads, and certain flatlands. Thus, in its occupied regions, vast ground is left unoccupied, which gives Chinese guerrilla war a vast area of activity. Nationwide, even if the enemy can occupy the line of Guangzhou, Wuhan, Lanzhou and nearby regions, regions beyond this are difficult to occupy; this gives China a general rear area and central base area for conducting protracted war and striving for final victory. Second, the enemy, with few soldiers confronting many soldiers, finds itself surrounded by many soldiers. The enemy advances toward us in separate columns; the enemy occupies strategic exterior lines, we occupy strategic interior lines; the enemy is on strategic offensive, we are on strategic defense; it appears I am very unfavorable. However, I can utilize the two advantages of vast territory and many soldiers, not adopting stubborn positional war, but adopting flexible mobile war, using several divisions against one enemy division, tens of thousands of men against ten thousand men, several routes against one route, suddenly surrounding one route from the exterior lines of the battlefield and attacking it. Thus, the enemy's exterior lines and offensive in strategic operations 不得不 (must) become interior lines and defense in campaign and battle operations. My interior lines and defense in strategic operations become exterior lines and offensive in campaign and battle operations. This applies to one route; it applies to other routes as well. The above two points both arise from the characteristic of enemy smallness and our largeness. Also, although enemy soldiers are few, they are strong soldiers (in terms of weapons and personnel training); although our soldiers are many, they are weak soldiers (also only in terms of weapons and personnel training, not morale); therefore, in campaign and battle operations, not only should I use many soldiers to fight few soldiers, fighting interior lines from exterior lines, but I must also adopt the policy of quick-decision war. To implement quick decision, generally do not fight stationary enemies, but fight enemies in motion. I will secretly concentrate large forces on the side of the enemy's necessary route; taking advantage of the enemy's movement, suddenly advance, surround and attack them, catching them off guard and resolving the battle quickly. If fought well, it is possible to eliminate all, most, or part of them; if fought poorly, it also inflicts great casualties on them. One battle is like this; all other battles are the same. To say no more, fighting one relatively large victorious battle every month, like Pingxingguan or Taierzhuang, can greatly depress the enemy's spirit, raise our army's morale, and summon world sympathy. Thus, my strategic protracted war becomes quick-decision war in battlefield operations. The enemy's strategic quick-decision war, after many defeated battles in campaigns and battles, 不得不 (must) change to protracted war.

(75) The above campaign and battle operational policy, to say it in one word, is: "*Exterior-line quick-decision offensive war.*" For my strategic policy of "interior-line protracted defensive war," it is opposite; however, it is precisely the necessary policy to implement such a strategic policy. If the campaign and battle policy were also "interior-line protracted defensive war," as was done in the early period of the War of Resistance, then it would be completely unsuitable for the two situations of enemy smallness and our largeness, enemy strength and our weakness; then it would absolutely not achieve the strategic purpose, not achieve total protracted war, and will be defeated by the enemy. Therefore, we have always advocated forming several large field army groups nationwide, their military strength targeting each of the enemy's field army groups' strength and doubling, tripling, or quadrupling it, adopting the above policy, maneuvering with the enemy on the vast battlefield. This policy is not only useful for regular war but also useful for guerrilla war, and must

be used. It is not only applicable to a certain stage of war but applicable to the entire process of war. In the strategic counteroffensive stage, my technical conditions strengthen; even if the situation of using weakness against strength is completely gone, I still use many soldiers from exterior lines to adopt quick-decision offensive war, which can achieve even greater results in capturing large quantities of prisoners. For example, if I use two or three or four mechanized divisions against one enemy mechanized division, I can more certainly eliminate this division. Several big men fighting one big man is easy to win; this is a truth contained in common sense.

(76) If we resolutely adopt *"exterior-line quick-decision offensive war"* in battlefield operations, not only will we change the situation of strength and weakness, advantage and disadvantage between enemy and ourselves on the battlefield, but we will also gradually change the general situation. On the battlefield, because I am on offensive, the enemy is on defense; I am many soldiers on exterior lines, the enemy is few soldiers on interior lines; I am quick-decision, the enemy although attempting to wait for reinforcement through protraction, cannot have his own way; thus on the enemy side, the strong becomes weak, advantage becomes disadvantage; on our army side, conversely, the weak becomes strong, disadvantage becomes advantage. After fighting many such victorious battles, the general enemy-versus-us situation will undergo change. That is to say, after assembling many victories of exterior-line quick-decision offensive war in battlefield operations, we gradually strengthen ourselves and weaken the enemy; thus the general situation of strength, weakness, advantage, and disadvantage cannot but be influenced and change. At that time, coordinating with our other conditions, and coordinating with internal changes within the enemy and favorable international situation, we can make the general enemy-versus-us situation walk to balance, then from balance to my advantage and enemy disadvantage. At that time, it is the opportunity for us to implement counteroffensive and drive the enemy out of the country.

(77) War is a contest of strength, but strength changes its original form in the process of war. Here, subjective effort, fighting more victorious battles and making fewer errors, is the decisive factor. Objective factors possess the possibility of this change, but to realize this possibility requires correct policy and subjective effort. At this time, *the subjective role is decisive.*

IX. Initiative, Flexibility, Planning; Forms of Warfare

A. Initiative, Flexibility, and Planning (§§78–90)

(78) The campaign and battle-level exterior-line quick-decision offensive war discussed above centers on *offense*; "exterior lines" denotes the spatial scope of the offensive, "quick decision" denotes its temporal dimension—hence the designation "exterior-line quick-decision offensive war." This constitutes the optimal policy for implementing protracted war, synonymous with what is termed *mobile warfare*. However, the execution of this policy is inseparable from *initiative, flexibility, and planning*. We now turn to examine these three questions.

(79) Having already discussed conscious initiative, why now speak of initiative? Conscious initiative refers to conscious activity and effort—the characteristic distinguishing humans from objects—which manifests with particular intensity in warfare, as previously stated. The *initiative* discussed here refers to the army's freedom of action, distinguishing it from a state of compelled passivity. Freedom of action is the lifeblood of an army; deprived of this freedom, an army approaches defeat or annihilation. A soldier being disarmed results from losing freedom of action and being forced into a passive position. The defeat of an army follows the same principle. For this reason, both sides in war strive for initiative and strive to avoid passivity. The exterior-line quick-decision offensive war we propose, along with the flexibility and planning required to implement it, may be said to aim at seizing initiative, compelling the enemy into passivity, and thereby achieving the purpose of preserving oneself and eliminating the enemy. However, initiative or passivity is inseparable from the superiority or inferiority of war strength, and thus also inseparable from the correctness or incorrectness of subjective guidance. Additionally, there exist situations in which one may seize initiative and compel enemy passivity by exploiting enemy misperceptions and unpreparedness. We now proceed to analyze these points.

(80) Initiative is inseparable from superiority in war strength, while passivity is inseparable from inferiority in war strength. Superiority or inferiority in war strength constitutes the objective basis for initiative or passivity. A strategically active position is naturally more readily mastered and exercised through strategic offensive operations; however, an active position that is thorough and pervasive—that is, absolute initiative—is possible only when absolute superiority confronts absolute inferiority.

When a physically robust person contends with a critically ill patient, the former possesses absolute initiative. If Japan did not possess many insurmountable contradictions—for example, if it could deploy several million to ten million troops at once, if its financial resources were several times greater than at present, if it faced no domestic or foreign opposition, and if it did not adopt barbarous policies provoking the Chinese people to resist unto death—then it could maintain a kind of absolute superiority and thereby possess an absolute initiative that is thorough and pervasive. However, historically, such matters of absolute superiority exist in the outcomes of wars and campaigns, but are rare at their outset. For example, on the eve of Germany's capitulation in the First World War, the Allied Powers had become absolutely superior while Germany had become absolutely inferior; the result was German defeat and Allied victory—an example of absolute superiority and inferiority existing in the outcome of a war. Similarly, on the eve of the victory at Taierzhuang, the isolated Japanese forces in that locality, after bitter fighting, had reached a state of absolute inferiority while our forces had achieved absolute superiority; the result was enemy defeat and our victory—an example of absolute superiority and inferiority existing in the outcome of a campaign. Wars or campaigns may also conclude with relative superiority/inferiority or a state of equilibrium; in such cases, war ends in compromise, while a campaign ends in stalemate. However, most are decided by absolute superiority/inferiority. All these pertain to the outcomes, not the beginnings, of wars or campaigns. The final outcome of the Sino-Japanese War can be foreseen: Japan will fail with absolute inferiority, while China will prevail with absolute superiority. However, at present, the superiority or inferiority of both sides is not absolute but relative. Japan, by virtue of its favorable factors—superior military strength, economic capacity, and political-organizational power—has gained advantage over our inferior military, economic, and political-organizational capacities, thereby laying the foundation for its initiative. However, because its military and other strengths are limited in quantity, and because it possesses many other unfavorable factors, its advantage is diminished by its own contradictions. Upon encountering China, it then confronts China's vast territory, large population, numerous troops, and resolute national resistance; its advantage is further diminished. Consequently, in the overall aspect, its position becomes a kind of relative superiority, and thus the exercise and maintenance of its initiative are constrained and become relative as well. On China's side, although inferior in the intensity of strength—thereby creating a certain strategic passivity—it enjoys superiority in geographical extent, population, and troop numbers, as well as in the fighting spirit and morale of its people and soldiers. This superiority, combined with other favorable factors, diminishes the degree of inferiority in our military and economic capacities, transforming it into a state of relative strategic inferiority. Consequently, the degree of passivity is also reduced, placing us in a position of merely relative strategic passivity. However, passivity is always disadvantageous and must be actively escaped. The military method is to resolutely implement exterior-line quick-decision offensive warfare and to vigorously develop guerrilla warfare behind enemy lines, thereby achieving many localized positions of superiority and initiative over the enemy through mobile and guerrilla warfare at the campaign and battle levels. Through many such localized campaign-level advantages and positions of initiative, one can gradually create strategic superiority and strategic initiative, thereby escaping strategic inferiority and passivity. Such is the mutual relationship between initiative and passivity, superiority and inferiority.

(81) From this, one may also understand the relationship between initiative or passivity and subjective guidance. As stated above, our relative strategic inferiority and strategic passivity can be escaped; the method is to artificially create many localized advantages and positions of initiative for ourselves, thereby depriving the enemy of many localized advantages and positions of initiative, casting him into inferiority and passivity. Aggregating these localized elements yields our strategic superiority and strategic initiative, and the enemy's strategic inferiority and strategic passivity. Such transformation relies upon correct subjective guidance. Why? Both I and the enemy desire advantage and initiative; from this perspective, war is a contest of subjective capabilities between the commanders of two armies, using military and financial strength and other material foundations as their stage, each striving for superiority and initiative. The outcome of the contest yields victory or defeat; besides the comparison of objective material conditions, the victor must owe his success to correct subjective command, while the vanquished must owe his failure to erroneous subjective command. We acknowledge that the phenomena of war are more elusive, less certain—that is, more "probabilistic"—than any other social phenomena. However, war is not a divine entity; it remains a kind of necessary movement in the world. Therefore, Sun Tzu's principle, "Know the enemy and know yourself; in a hundred battles, you will never be in peril" [24], remains a scientific truth.

Errors arise from ignorance of the enemy and oneself; the nature of war also makes it impossible in many situations to fully know both, thereby generating uncertainty in war conditions and war actions, producing errors and failures. However, regardless of war conditions and actions, knowing their general outline and essential points is possible. Through various reconnaissance methods followed by the commander's intelligent inference and judgment, reducing errors and achieving generally correct guidance is attainable. Armed with this "generally correct guidance" as a weapon, one can win more battles, transform inferiority into superiority, and passivity into initiative. Such is the relationship between initiative or passivity and the correctness or incorrectness of subjective guidance.

(82) The correctness or incorrectness of subjective guidance affects the transformation of superiority/inferiority and initiative/passivity; this is further confirmed by historical facts of powerful armies suffering defeat and weak armies achieving victory. There are many such instances in Chinese and foreign history. In China: the Battle of Chengpu between Jin and Chu [25], the Battle of Chenggao between Chu and Han [26], Han Xin's defeat of Zhao [27], the Battle of Kunyang between Xin and Han [28], the Battle of Guandu between Yuan and Cao [29], the Battle of Red Cliffs between Wu and Wei [30], the Battle of Yiling between Wu and Shu [31], the Battle of Fei River between Qin and Jin [32], and so forth. Abroad: most of Napoleon's campaigns [33], the Soviet civil war following the October Revolution—all were cases of defeating the many with the few, achieving victory despite inferiority confronting superiority. In each case, one first achieved localized superiority and initiative against the enemy's localized inferiority and passivity, won victory in one engagement, then extended to the remainder, defeating them one by one, thereby transforming the overall situation into superiority and initiative. For the enemy originally possessing superiority and initiative, the opposite occurs; due to subjective errors and internal contradictions, his excellent or relatively good position of superiority and initiative can be entirely lost, transforming him into a defeated general or a ruined monarch. From this, it is evident that while the superiority or inferiority of war strength itself constitutes the objective basis for determining initiative or passivity, it is not yet the actual reality of initiative or passivity; only after struggle, after the contest of subjective capabilities, does factual initiative or passivity emerge. In the struggle, due to the correctness or incorrectness of subjective guidance, one may transform inferiority into superiority, passivity into initiative; or conversely, transform superiority into

inferiority, initiative into passivity. That no ruling dynasty can defeat a revolutionary army demonstrates that mere superiority has not yet determined a position of initiative, let alone final victory. Initiative and victory can be seized from the hands of those possessing superiority and initiative by those in inferiority and passivity, based on actual conditions, through the activation of subjective capabilities, and by attaining certain conditions.

(83) Misperception and unpreparedness can cause loss of superiority and initiative. Therefore, deliberately creating enemy misperceptions and launching unexpected attacks constitutes a method—and an important one—for creating superiority and seizing initiative. What is misperception? "On Mount Bagong, every bush and tree seemed a soldier" [34] is one example. "Feinting east while striking west" is one method for creating enemy misperception. When favorable popular conditions exist, sufficient to seal off information, employing various methods of deceiving the enemy can often effectively trap the enemy in the bitter predicament of erroneous judgment and erroneous action, thereby causing loss of his superiority and initiative. "In war, deception is never exhausted" refers precisely to this matter. What is unpreparedness? It is lack of readiness. Superiority without preparedness is not true superiority, nor does it confer initiative. Understanding this point, an inferior but prepared force can often launch an unexpected offensive against the enemy, defeating the superior party. We say that a moving enemy is easier to engage precisely because the enemy is in a state of unpreparedness—that is, unready. These two matters—creating enemy misperceptions and launching unexpected attacks—mean imparting uncertainty to the enemy through the uncertainty inherent in war, while securing for oneself the greatest possible certainty, thereby striving for one's own superiority and initiative, striving for one's own victory. To achieve these, the prerequisite condition is superior popular organization. Therefore, mobilizing all people opposed to the enemy, arming them uniformly, conducting extensive raids against the enemy, and simultaneously using them to seal off information and shield our forces—so that the enemy cannot know where or when our forces will attack him, thereby creating the objective basis for his misperception and unpreparedness—is extremely important. In the past, during the era of the agrarian revolutionary war, the Chinese Red Army, with its weak military strength, frequently won victories; this owed much to the organization and arming of the masses. By rights, a national war should be able to secure even broader popular support than an agrarian revolutionary war; however, due to historical errors [35], the masses are dispersed, not only difficult to utilize hastily but also at times exploited by the enemy. Only by resolutely and extensively mobilizing the entire populace can one provide inexhaustible supply for all needs of the war. In this method of warfare—imparting misperception and unpreparedness to the enemy in order to defeat him—it will certainly play a major role. We are not Duke Xiang of Song; we reject that swinish "benevolence and righteousness" [36]. We must seal the enemy's eyes and ears as much as possible, turning them into the blind and deaf; we must sow as much confusion as possible in the minds of their commanders, turning them into madmen, thereby striving for our own victory. All these also constitute the mutual relationship between initiative or passivity and subjective guidance. Such subjective guidance is indispensable for defeating Japan.

(84) Broadly speaking, during its offensive phase, Japan generally occupies a position of initiative due to its military strength and its exploitation of our historical and current subjective errors. However, this initiative has already begun to weaken in part, owing to the many unfavorable factors it carries, the subjective errors it has committed in the course of the war (discussed in detail later), and the many favorable factors possessed by our side. The enemy's defeat at Taierzhuang and its predicament in Shanxi are clear evidence. The extensive development of our guerrilla warfare behind enemy lines has placed the enemy's garrison forces in its occupied territories in a completely passive position. Although the enemy remains engaged in its active strategic offensive at present, its initiative will conclude with the cessation of its strategic offensive. The enemy's insufficiency of troops, which precludes unlimited offensive operations, is the first root cause of its inability to maintain initiative. Our campaign-level offensive warfare, guerrilla warfare behind enemy lines, and other conditions constitute the second root cause compelling it to halt its offensive at a certain limit and preventing it from maintaining initiative. The existence of the Soviet Union and other international changes constitute the third root cause. From this, it is evident that the enemy's position of initiative is limited and can be undermined. If China can persist in the campaign and battle-level offensive warfare of its main forces in its methods of warfare, vigorously develop guerrilla warfare behind enemy lines, and extensively mobilize the masses politically, our strategic position of initiative can gradually be established.

(85) We now turn to flexibility. What is flexibility? It is the concrete realization of initiative in warfare; it is the flexible employment of forces. The flexible employment of forces is the central task of war command and also the most difficult to execute well. The enterprise of war, besides organizing and educating the army and organizing and educating the people, consists in employing the army in combat—and everything is for the sake of victory in combat. Organizing the army and so forth is certainly difficult, but employing the army is even more difficult, especially in situations of confronting strength with weakness. Executing this task requires immense subjective capability; it requires overcoming the confusion, darkness, and uncertainty inherent in the nature of war, and discovering order, clarity, and certainty therein, before achieving flexibility in command.

(86) The basic policy for battlefield operations in the War of Resistance Against Japan is exterior-line quick-decision offensive warfare. Implementing this policy involves various tactics or methods: dispersal and concentration of forces, separate advance and convergent attack, offense and defense, assault and containment, encirclement and outflanking, advance and retreat. Understanding these tactics is easy; flexibly employing and transforming them is not easy. Here there are three critical junctures: timing, location, and unit. Without the proper timing, without the proper location, without suitability to the condition of the unit, victory cannot be achieved. For example, when attacking a moving enemy, striking too early exposes oneself and gives the enemy conditions for preparation; striking too late allows the enemy to concentrate and halt, becoming a hard nut to crack. This is the question of timing. Selecting the point of assault on the left flank, where it strikes the enemy's weakness, facilitates victory; selecting the right flank, where it encounters the enemy's strongpoint, proves ineffective. This is the question of location. Executing a certain task with one of our units facilitates victory; executing the same task with another unit proves difficult. This is the question of the unit's condition. Not only must tactics be employed, but they must also be transformed. Offense becomes defense, defense becomes offense, advance becomes retreat, retreat becomes advance, containment units become assault units, assault units become containment units, and encirclement and outflanking mutually transform—based on the conditions of enemy and our forces, enemy and our terrain, transforming them timely and appropriately is an important task of flexible command. This applies to battle command; it also applies to campaign and strategic command.

(87) The ancients said, "The marvel of employment lies in the mind" [37]. This "marvel" we term flexibility; it is the product of the intelligent commander. Flexibility is not reckless action; reckless action should be rejected. Flexibility is the capability of the intelligent commander, based on objective conditions, to "assess the times and size up the situation" (this "situation" includes the enemy's situation, our situation, terrain, and so forth) and to adopt timely and appropriate methods of disposition—that is, the so-called "marvel of employment." Based on this marvel of employment, exterior-line quick-decision offensive warfare can achieve victory more frequently, can transform the situation of superiority and inferiority between enemy and ourselves, can realize our initiative over the enemy, can overwhelm and defeat the enemy, and final victory will belong to us.

(88) We now turn to planning. Due to the uncertainty peculiar to war, realizing planning in war is far more difficult than realizing planning in other enterprises. However, "In all things, preparation leads to success; lack of preparation leads to failure" [38]; without prior planning and preparation, victory in war cannot be achieved. War possesses no absolute certainty, but it is not devoid of a certain degree of relative certainty. Our own side is relatively certain. The enemy's side is highly uncertain, but there are signs to be discerned, indications to be observed, preceding and succeeding phenomena to be considered. This constitutes the so-called certain degree of relative certainty, providing an objective basis for planning in war. The development of modern technology (wired and wireless telegraphy, aircraft, automobiles, railways, steamships, etc.) has further increased the possibility of planning in war. However, because war possesses only a relatively low degree and brief duration of certainty, planning in war can hardly be complete and fixed; it moves with the movement (or flow, or progression) of war, and varies in degree according to the scope of the war. Tactical plans—for example, attack or defense plans for small units and small forces—often require modification several times daily. Campaign plans—that is, operational plans for large formations—can generally endure until the conclusion of the campaign, but within that campaign, partial changes are common, and complete changes occasionally occur. Strategic plans, based on the overall situation of both sides in the war, possess a greater degree of fixity, but apply only within a certain strategic stage; when war progresses to a new stage, strategic plans must be altered. The determination and alteration of tactical, campaign, and strategic plans according to their respective scope and conditions constitutes an important juncture of war command; it is also the concrete implementation of flexibility in war, and also the actual "marvel of employment." Commanders at all levels in the War of Resistance Against Japan should pay attention to this.

(89) Some people, based on the fluidity of war, fundamentally deny the relative fixity of war plans or war policies, claiming that such plans or policies are "mechanical." This view is erroneous. As stated in the preceding section, we fully acknowledge: because war conditions possess only relative certainty and war rapidly flows forward (or moves, progresses), war plans or policies should only be accorded relative fixity and must be replaced or revised in a timely manner according to changes in conditions and the fluidity of war; failing to do so would render us mechanical. However, one absolutely cannot deny the relatively fixed war plan or policy applicable within a certain period of time; denying this point denies everything, including war itself and the person speaking. Because the conditions and actions of war possess their relative fixity, the war plans or policies arising in response must also be endowed with relative fixity. For example, because the conditions of the war in North China and the dispersed operations of the Eighth Route Army possess fixity within a certain stage, it is entirely necessary to endow relative fixity upon the Eighth Route Army's strategic operational policy of "guerrilla warfare as fundamental, but not relaxing mobile warfare under favorable conditions" within that certain stage. Campaign policies, compared to the aforementioned strategic policies, apply for shorter durations; tactical policies are even shorter; however, all possess a certain degree of fixity for a certain period of time. Denying this point renders war impossible to undertake, becoming devoid of definite views—this is not, that is not, or this is, that is—the relativism of war. No one denies that even a policy applicable within a certain definite period of time is also in flux; without this flux, there would be no abandonment of one policy and adoption of another. However, this flux is limited—it flows within the scope of various different war actions executing this policy, not a flux in the fundamental nature of this policy; that is to say, it is a flux of quantity, not of quality. This fundamental nature, within a certain period of time, absolutely does not flux; what we term relative fixity within a certain period of time refers precisely to this point. In the absolutely fluid entire river of war, there exists relative fixity at each specific stage—such is our view of the fundamental nature of war plans or war policies.

(90) Having discussed strategic interior-line protracted defensive warfare and campaign/battle-level exterior-line quick-decision offensive warfare, and having discussed initiative, flexibility, and planning, we may now summarize in a few words. The War of Resistance Against Japan should be planned. War plans—that is, the concrete application of strategy and tactics—should embody flexibility, enabling adaptation to war conditions. Everywhere, attention should be paid to transforming inferiority into superiority, passivity into initiative, in order to change the situation between enemy and ourselves. All these manifest in campaign and battle-level exterior-line quick-decision offensive warfare, and simultaneously manifest in strategic interior-line protracted defensive warfare.

B. Mobile Warfare, Guerrilla Warfare, Positional Warfare (§§91–96)

(91) The campaign and battle-level exterior-line quick-decision offensive warfare within strategic interior lines, strategic protraction, and strategic defense—when expressed in the form of warfare—manifests as *mobile warfare*. Mobile warfare is the form wherein regular forces engage in campaign and battle-level exterior-line quick-decision offensive warfare along extended fronts and in vast war zones. Simultaneously, it includes the so-called "mobile defense" executed at certain necessary moments to facilitate the execution of such offensive warfare, and also includes positional attack and positional defense playing an auxiliary role. Its characteristics are: regular forces, superiority of forces at the campaign and battle levels, offensiveness, and mobility.

(92) China possesses a vast territory and numerous troops, but its army's technical conditions and training are inadequate; the enemy, by contrast, has insufficient troops but relatively superior technical conditions and training. Under such circumstances, offensive mobile warfare should undoubtedly constitute the primary form of warfare, with other forms playing an auxiliary role, composing the entirety of mobile warfare. Here, one must oppose the so-called "retreat without advance" escapism, and simultaneously oppose the so-called "advance without retreat" recklessness.

(93) One characteristic of mobile warfare is its mobility, which not only permits but demands large-scale advances and retreats by field armies. However, this has nothing in common with the escapism of the Han Fuju type [39]. The basic requirement of war is: eliminate the enemy; its other requirement is: preserve oneself. The purpose of preserving oneself lies in eliminating the enemy; and eliminating the enemy is, in turn, the most effective means of preserving oneself. Therefore, mobile warfare must never be used as a pretext by Han Fuju-type individuals; it is by no means only backward movement without forward movement; such "movement" negates the basic offensiveness of mobile warfare, and its implementation would result in China, though vast, being "moved" away entirely.

(94) However, another line of thought is also incorrect: the so-called "advance without retreat" recklessness. We advocate mobile warfare whose content consists of campaign and battle-level exterior-line quick-decision offensive warfare; this includes positional warfare playing an auxiliary role, and also includes "mobile defense" and retreat; without these, mobile warfare cannot be fully executed. Recklessness is military myopia; its root often lies in fear of losing territory. The reckless do not understand that one characteristic of mobile warfare is its mobility, which not only permits but demands large-scale advances and retreats by field armies. Positively, in order to place the enemy in disadvantageous conditions favorable to our operations, it is often necessary to require the enemy to be in motion and to secure many conditions favorable to us—for example, advantageous terrain, favorable enemy dispositions, residents capable of sealing off information, enemy fatigue and unpreparedness, etc. This demands the enemy's advance, even at the cost of temporarily losing part of our territory. Because temporarily and partially losing territory is the price for wholly and permanently preserving and recovering territory. Negatively, whenever compelled into a disadvantageous position fundamentally threatening the preservation of military strength, one should courageously retreat in order to preserve military strength and strike the enemy again under new conditions. The reckless do not understand this principle; clearly already in a confirmed disadvantageous situation, they still contend over the gain or loss of a city or plot of land, with the result that not only are city and land both lost, but military strength cannot be preserved either. We have always advocated "luring the enemy in deep" precisely because this constitutes the most effective military policy for a weak force confronting a strong force in strategic defense.

(95) Among the forms of warfare in the War of Resistance Against Japan, the primary is mobile warfare; the next is guerrilla warfare. We say that, in the war as a whole, mobile warfare is primary and guerrilla warfare auxiliary, meaning that resolving the fate of the war relies primarily on regular warfare, especially the mobile warfare therein; guerrilla warfare

cannot bear the primary responsibility for resolving the war's fate. However, this does not mean that guerrilla warfare's strategic position in the War of Resistance Against Japan is unimportant. The strategic position of guerrilla warfare in the entire War of Resistance Against Japan is second only to mobile warfare; for without the assistance of guerrilla warfare, the enemy cannot be defeated. Stated thus, this includes the strategic task of guerrilla warfare developing into mobile warfare. During the long and cruel war, guerrilla warfare will not remain in its original position; it will elevate itself to mobile warfare. Thus, the strategic function of guerrilla warfare has two aspects: first, assisting regular warfare; second, transforming itself into regular warfare. Regarding the unprecedentedly extensive and unprecedentedly protracted significance of guerrilla warfare in China's War of Resistance Against Japan, its strategic position is even more not to be slighted. Therefore, in China, guerrilla warfare itself involves not only tactical questions but also its special strategic questions. This question I have already addressed in my article "On the Strategic Questions of Anti-Japanese Guerrilla Warfare." As stated earlier, the forms of warfare in the three strategic stages of the War of Resistance Against Japan are: in the first stage, mobile warfare is primary, with guerrilla warfare and positional warfare auxiliary. In the second stage, guerrilla warfare will ascend to primary position, with mobile warfare and positional warfare auxiliary. In the third stage, mobile warfare will again ascend to primary form, with positional warfare and guerrilla warfare auxiliary. However, the mobile warfare of this third stage will not be borne entirely by the original regular forces; part of it—perhaps a quite important part—will be borne by the original guerrilla forces elevating themselves from guerrilla warfare to mobile warfare. Viewed across the three stages, guerrilla warfare in China's War of Resistance Against Japan is by no means dispensable. It will enact an unprecedentedly magnificent scene in the history of human warfare. For this reason, among the several million regular troops nationwide, designating at least several hundred thousand to disperse across all enemy-occupied regions, mobilizing and coordinating with popular armed forces, and engaging in guerrilla warfare, is entirely necessary. The designated forces should consciously undertake this sacred task; they should not think that fighting fewer large battles makes them appear less like national heroes for a time, diminishing their qualifications—such thinking is erroneous. Guerrilla warfare does not have the rapid efficacy and illustrious reputation of regular warfare, but "a long journey tests a horse's strength; time reveals a person's heart"; in the long and cruel war, guerrilla warfare will manifest its great power; it is truly no small undertaking. Moreover, regular forces dispersing to conduct guerrilla warfare can, when assembled, again conduct mobile warfare; the Eighth Route Army does precisely this. The policy of the Eighth Route Army is: "Guerrilla warfare is fundamental, but mobile warfare under favorable conditions must not be relaxed." This policy is entirely correct; the views of those who oppose this policy are incorrect.

(96) Defensive and offensive positional warfare, under China's present technical conditions, generally cannot be executed; this also constitutes a manifestation of our weakness. Furthermore, the enemy exploits China's vast territory to avoid our positional installations. Therefore, positional warfare cannot be employed as an important means, let alone as a primary means. However, during the first and second stages of the war, localized positional warfare included within the scope of mobile warfare and playing an auxiliary role in campaign operations is possible and necessary. Adopting semi-positional so-called "mobile defense" for the purpose of resisting step by step in order to exhaust the enemy and gain time for maneuver belongs even more to the necessary part of mobile warfare. China must strive to increase new-type weapons in order to be able to fully execute the task of positional attack during the strategic counteroffensive stage. The strategic counteroffensive stage will undoubtedly elevate the position of positional warfare, because at that time the enemy will hold positions firmly; without our powerful positional attacks coordinating with mobile warfare, the purpose of recovering lost territory cannot be achieved. Nevertheless, even in the third stage, we must still strive to make mobile warfare the primary form of war. Because the art of war leadership and human activity die by more than half when confronted with positional warfare of the type seen in Western Europe after the middle period of the First World War. However, conducting operations within China's vast territory, and for a considerable length of time China still retaining the condition of technical weakness, the matter of "liberating war from the trenches" naturally emerges. Even in the third stage, although China's technical conditions will have improved, they still may not surpass those of the enemy; thus one is compelled to strive for highly developed mobile warfare if the purpose of final victory is to be achieved. Thus, throughout the entire War of Resistance Against Japan, China will not take positional warfare as its primary form; the primary and important forms will be mobile warfare and guerrilla warfare. In these forms of warfare, the art of

war leadership and human activity can obtain full opportunity for expression; this is, again, our misfortune containing a fortunate element!

X. Annihilation vs. Attrition; Exploiting Vulnerabilities; Decisive Engagements; The People as the Foundation of Victory; Conclusion

A. Warfare of Annihilation and Warfare of Attrition (§§97–104)

(97) As stated earlier, the essence of war—that is, its purpose—is to preserve oneself and eliminate the enemy. However, the forms of warfare for achieving this purpose include mobile warfare, positional warfare, and guerrilla warfare; the effects realized in their execution differ in degree, and thus there is generally a distinction between *warfare of attrition and warfare of annihilation.*

(98) We may first state that the War of Resistance Against Japan is simultaneously a warfare of attrition and a warfare of annihilation. Why? The enemy's factors of strength are still being exerted; his strategic advantage and initiative remain intact. Without warfare of annihilation at the campaign and battle levels, it is impossible to effectively and rapidly diminish his factors of strength and undermine his advantage and initiative. Our factors of weakness also persist; our strategic disadvantage and passive position have not yet been escaped. In order to gain time, strengthen domestic and international conditions, and transform our unfavorable state, without warfare of annihilation at the campaign and battle levels, success cannot be achieved. Therefore, *warfare of annihilation at the campaign level is the means for achieving the strategic purpose of warfare of attrition.* In this respect, warfare of annihilation is warfare of attrition. China's ability to conduct a protracted war relies principally on the method of achieving attrition through annihilation.

(99) However, there is also warfare of attrition at the campaign level that serves the strategic purpose of attrition. Broadly speaking, mobile warfare is primarily for executing tasks of annihilation; positional warfare is primarily for executing tasks of attrition; guerrilla warfare executes tasks of both attrition and annihilation, with distinctions among the three. In this respect, warfare of annihilation differs from warfare of attrition. Warfare of attrition at the campaign level is auxiliary, yet it is also necessary for protracted warfare.

(100) Speaking theoretically and in terms of necessity, during China's strategic defensive stage, we should utilize the principal annihilative character of mobile warfare, the partial annihilative character of guerrilla warfare, plus the principal attritive character of auxiliary positional warfare and the partial attritive character of guerrilla warfare, to achieve the strategic purpose of extensively consuming the enemy. During the strategic stalemate stage, we continue to utilize the annihilative and attritive characters of guerrilla warfare and mobile warfare to extensively consume the enemy once again. All these serve to prolong the war situation, gradually transform the situation between enemy and ourselves, and prepare the conditions for counteroffensive. During the strategic counteroffensive stage, we continue to use annihilation to achieve attrition, in order to finally expel the enemy from the country.

(101) However, in actual fact, the experience of ten months shows that many, even the majority, of mobile warfare campaigns have been conducted as warfare of attrition; the due annihilative function of guerrilla warfare, in certain regions, has also not yet been elevated to the requisite degree. The advantage of this situation is that, in any case, we have at least consumed the enemy, which has significance for protracted warfare and ultimate victory—our blood has not been shed in vain. However, the disadvantages are: first, insufficient consumption of the enemy; second, our own forces inevitably suffer greater consumption while capturing fewer spoils. Although we should acknowledge the objective causes of this situation—namely, the differences in technical conditions and troop training between enemy and ourselves—nevertheless, both theoretically and practically, we should in every way advocate that the main forces strive to execute warfare of annihilation on all favorable occasions. Although guerrilla units, in order to execute many concrete tasks—such as sabotage and disruption—must conduct pure warfare of attrition, we must still advocate and strive to implement annihilative warfare on all favorable occasions at the campaign and battle levels, in order to achieve the purpose of both extensively consuming the enemy and extensively replenishing ourselves.

(102) The so-called "exterior lines," "quick decision," and "offensive" of exterior-line quick-decision offensive warfare, as well as the so-called "mobility" of mobile warfare, in terms of combat form, principally employ the tactics of encirclement and outflanking, and thus require the concentration of superior forces. Therefore, *the concentration of forces and the adoption of encirclement and outflanking tactics are necessary conditions for implementing mobile warfare—that is, exterior-line quick-*

decision offensive warfare. However, all these serve the purpose of annihilating the enemy.

(103) The strengths of the Japanese army lie not only in its weapons but also in the training of its officers and soldiers—its organizational character, its self-confidence formed from never having suffered defeat, its superstition regarding the Emperor and deities, its arrogance and self-esteem, its contempt for the Chinese people, and so forth. These characteristics have been created by years of arbitrary education by the Japanese military clique and by Japanese national habits. The phenomenon that our army inflicts heavy casualties but captures few prisoners principally stems from this point. Many people in the past underestimated this factor. The destruction of these characteristics requires a prolonged process. First, we must attach importance to this characteristic; then, patiently and systematically, from political, international propaganda, and Japanese popular movement perspectives, we must conduct work toward this end; warfare of annihilation on the military front is also one method. Here, pessimists may use this point to advance the theory of national subjugation, while passive military strategists may use it to oppose warfare of annihilation. We hold the opposite view: we believe that this strength of the Japanese army can be destroyed, and indeed has already begun to be destroyed. The principal method of destruction is political persuasion. Regarding Japanese soldiers, rather than insulting their self-esteem, we should understand and properly guide this self-esteem; through lenient treatment of prisoners of war, we should guide them to understand the anti-popular aggression of Japan's rulers. On the other hand, we should demonstrate before them the indomitable spirit and heroic, tenacious combat effectiveness of the Chinese army and people—this constitutes the strike of warfare of annihilation. Speaking operationally, the experience of ten months proves that annihilation is possible; the campaigns at Pingxingguan, Taierzhuang, and others are clear evidence. Japanese morale has already begun to waver; soldiers do not understand the purpose of the war; they find themselves surrounded by the Chinese army and people; their courage to charge is far weaker than that of Chinese soldiers—all these are objective conditions favorable to our conduct of warfare of annihilation, and these conditions will develop increasingly with the prolongation of the war. Speaking from the perspective of destroying the enemy's spirit through warfare of annihilation, annihilation is also one of the conditions for shortening the war process and hastening the liberation of Japanese soldiers and people. In the world, there are only cases of cats befriending cats; there are no cases of cats befriending mice [43].

(104) On the other hand, we should acknowledge that, in terms of technical conditions and the degree of troop training, we currently fall short of the enemy. Consequently, annihilation to the maximum extent—for example, capturing all or most of the enemy—is difficult in many situations, especially in battles on flat terrain. The excessive demands of proponents of quick victory on this point are also incorrect. The correct requirement for the War of Resistance Against Japan should be: *warfare of annihilation to the greatest extent possible*. On all favorable occasions, in every engagement, concentrate superior forces and adopt encirclement and outflanking tactics—if unable to encircle all, then encircle a part; if unable to capture all that is encircled, then capture a part; if unable to capture a part of what is encircled, then inflict heavy casualties on a part of what is encircled. On all occasions unfavorable to executing warfare of annihilation, conduct warfare of attrition. For the former, apply the principle of concentrating forces; for the latter, apply the principle of dispersing forces. In terms of campaign command relationships, for the former, apply the principle of centralized command; for the latter, apply the principle of decentralized command. These constitute the basic policy for battlefield operations in the War of Resistance Against Japan.

B. The Possibility of Exploiting Enemy Vulnerabilities (§105)

(105) Regarding the enemy's vulnerability to defeat, there is also a basis in the enemy's command. Since ancient times, there has never been a general who never commits errors; the enemy has weaknesses to be found, just as we ourselves are inevitably prone to errors—*the possibility of exploiting the enemy's vulnerabilities exists*. Speaking strategically and operationally, during the ten months of aggressive warfare, the enemy has already committed many errors. Counting the major ones, there are five:

First, *gradual increase of forces*. This stems from the enemy's underestimation of China, and also from his own insufficiency of forces. The enemy has always looked down upon us; after gaining advantages in the three northeastern provinces [40], plus the occupation of eastern Hebei and northern Chahar, all these counted as the enemy's strategic reconnaissance. Their conclusion was: "a loose sheet of sand." On this basis, they deemed China not worth fighting, and formulated the so-called "quick decision" plan, deploying minimal forces in an attempt to intimidate us into collapse. Over ten months, they did not anticipate China's such great unity and such great resistance; they forgot that China has entered a progressive era, that advanced political parties, advanced armies, and an advanced populace exist in China. When their plan failed, they gradually increased forces, from a dozen divisions repeatedly increased to thirty. To advance further, further increases are necessary. However, due to opposition from the Soviet Union, plus inherent insufficiency of personnel and finances, Japan's maximum troop deployment and final offensive points have had to be subject to certain limitations.

Second, *no main direction of attack*. Before the Taierzhuang campaign, the enemy broadly divided forces equally between Central and North China, and within each theater, forces were again divided equally. For example, in North China, forces were equally divided among the Tianjin-Pukou, Beiping-Hankou, and Datong-Puzhou railways; after suffering casualties on each route and garrisoning occupied territories, no forces remained for further advance. After the defeat at Taierzhuang, lessons were drawn, and main forces were concentrated in the Xuzhou direction—this error was temporarily corrected.

Third, *lack of strategic coordination*. Between the enemy's Central and North China theater groups, coordination within each group was broadly maintained, but coordination between the two groups was very poor. When the southern section of the Tianjin-Pukou Railway fought at Xiaobengbu, the northern section did not move; when the northern section fought at Taierzhuang, the southern section did not move. After both locations encountered setbacks, the Minister of War came to inspect, the Chief of the General Staff came to command—coordination was temporarily achieved. The rather serious contradictions existing within the Japanese landlord bourgeoisie and military clique are developing forward; lack of coordination in warfare is one concrete manifestation.

Fourth, *loss of strategic opportunity*. This is notably manifested in the pauses after the occupation of Nanjing and Taiyuan, principally due to insufficiency of forces and lack of strategic pursuit units.

Fifth, *much encirclement, little annihilation*. Before the Taierzhuang campaign, in the battles of Shanghai, Nanjing, Cangzhou, Baoding, Nankou, Xinkou, Linfen, and others, the enemy broke many forces but captured few prisoners, manifesting the clumsiness of his command.

These five—gradual increase of forces, no main direction of attack, lack of strategic coordination, loss of strategic opportunity, much encirclement with little annihilation—were the points of inadequacy in Japanese command before the Taierzhuang campaign. After Taierzhuang, although some corrections were made, given factors such as insufficiency of forces and internal contradictions, it is impossible to avoid repeating errors. Moreover, gains in one area entail losses in another: for example, concentrating North China forces on Xuzhou left large gaps in North China's occupied territories, providing guerrilla warfare with opportunities for unrestrained development. The above are errors the enemy made on his own, not errors we caused.

On our side, we may still intentionally create enemy errors—that is, through our own intelligent and effective actions, under the cover of organized popular forces, create enemy misperceptions, maneuver the enemy into our desired scope of action, such as "feinting east while striking west"—the possibility of this was discussed earlier. All these demonstrate: *the victory of our war may also find certain roots in the enemy's command*. Although we should not take this point as an important basis for our strategic plans—on the contrary, it is a reliable approach to base our plans on the assumption that the enemy commits few errors— nevertheless, just as we may exploit the enemy's vulnerabilities, the enemy may also exploit ours; minimizing the vulnerabilities we expose to the enemy is also a task of our command. However, the enemy's command errors are facts that have already existed, will occur again, and may be manufactured through our efforts—all these are sufficient for our utilization, and anti-Japanese generals should strive to seize them. Although the enemy's strategic and operational command is often inadequate, his combat command—that is, unit tactics and small-unit tactics—possesses considerable excellence; this is a point from which we should learn.

C. The Question of Decisive Engagements in the War of Resistance (§§106–110)

(106) The question of decisive engagements in the War of Resistance Against Japan should be divided into three categories:

1. All campaigns and battles with assurance of victory should be resolutely conducted as decisive engagements.
2. All campaigns and battles without assurance of victory should avoid decisive engagements.
3. Strategic decisive engagements that gamble the fate of the nation should be fundamentally avoided.

The War of Resistance Against Japan differs from many other wars precisely in this question of decisive engagements. During the first and second stages, with the enemy strong and we weak, the enemy's demand is that we concentrate main forces to engage in decisive battle with him. Our demand is the opposite: select favorable conditions, concentrate superior forces, and conduct decisive engagements at the campaign and battle levels with assurance of victory—such as Pingxingguan, Taierzhuang, and many other battles; while avoiding decisive engagements under unfavorable conditions without assurance of victory— such as the policy adopted in campaigns at Zhangde and other locations. *Strategic decisive engagements that gamble the fate of the nation are fundamentally not to be undertaken*—such as the recent withdrawal from Xuzhou. In this way, we undermine the enemy's "quick decision" plan and compel him to follow us in a protracted war. This policy is impossible to implement in countries with narrow territory, and difficult to implement in countries with excessively backward politics. We are a large country in a progressive era; this is achievable. If strategic decisive engagements are avoided, "so long as the green hills remain, there will be no shortage of firewood" [44]; although some territory is lost, there remains extensive room for maneuver, which can promote and await domestic progress, international reinforcement, and enemy internal collapse—this is the superior strategy for the War of Resistance Against Japan. Proponents of impetuous quick victory cannot endure the difficult journey of protracted war; seeking swift victory, they blow the trumpet for strategic decisive engagements as soon as the situation improves somewhat. If acted upon, the entire War of Resistance would suffer great loss, protracted war would be sacrificed, and precisely the enemy's poisonous scheme would be realized—this is truly the inferior strategy. Not to engage in decisive battle necessitates abandoning territory; this is beyond question. Under unavoidable circumstances (and only under such circumstances), one must courageously abandon territory. When the situation reaches this point, there should not be the slightest hesitation; this is the correct policy of exchanging territory for time. Historically, Russia avoided decisive engagements, executed courageous retreats, and defeated the once-mighty Napoleon [41]. China should now do the same.

(107) "Are we not afraid of being scolded for 'non-resistance'?" We are not afraid. Fundamentally not fighting and compromising with the enemy is the policy of non-resistance; this not only deserves condemnation but is completely impermissible. Resolutely resisting while avoiding the enemy's poisonous schemes, preventing our main forces from being annihilated by a single enemy strike—which would affect the continuation of the War of Resistance—in a word, avoiding national subjugation, is completely necessary. To harbor doubts on this point is myopia on the question of war; the result will inevitably be alignment with proponents of the theory of national subjugation. *We have criticized the so-called "advance without retreat" recklessness* precisely because if such recklessness becomes a general trend, the result may render the War of Resistance unable to continue, ultimately leading to the danger of national subjugation.

(108) We advocate decisive engagements under all favorable conditions, whether at the battle level or at the campaign level, large or small; on this point, no passivity is permissible. Inflicting annihilation or attrition upon the enemy can only be achieved through such decisive engagements; every anti-Japanese soldier must resolutely carry this out. For this purpose, partial and quite substantial sacrifices are necessary; the view that avoids any sacrifice is the view of cowards and those afflicted with fear of Japan, and must be resolutely opposed. The execution of deserters such as Li Fuying and Han Fuju [39] was correct. Promoting the spirit and actions of brave sacrifice and heroic advance in warfare, under correct operational plans, is absolutely necessary and inseparable from protracted war and ultimate victory. *We have severely condemned the so-called "retreat without advance" escapism* and advocated strict discipline precisely because only heroic decisive engagements under correct plans can defeat a strong enemy; whereas escapism is the direct supporter of the theory of national subjugation.

(109) "Is there not a contradiction in fighting bravely first and then abandoning territory afterward? Has not the blood of these brave fighters been shed in vain?" This is a highly inappropriate question. "Is eating first and defecating afterward eating in vain? Is sleeping first and waking afterward sleeping in vain?" Can one pose questions in this manner? I think not. Eating continuously without stopping, sleeping continuously without waking, fighting bravely continuously until reaching the Yalu River—this is the fantasy of subjectivism and formalism; it does not exist in actual life. Who does not know that, in fighting and shedding blood to gain time and prepare for counteroffensive, although certain territory may still inevitably be abandoned, time has been gained, the purpose of inflicting annihilation and attrition upon the enemy has been achieved, one's own combat experience has been acquired, the previously unmobilized populace has been mobilized, and international standing has increased? Has this blood been shed in vain? Not in the least. *Abandoning territory is for the purpose of preserving military strength, and precisely for the purpose of preserving territory*; because if one does not abandon partial territory under unfavorable conditions and blindly conducts absolutely unassured decisive engagements, the result—after losing military strength—will inevitably be followed by loss of all territory, let alone any hope of recovering lost territory. A capitalist conducting business must have capital; after complete bankruptcy, one is no longer a capitalist. A gambler must have stakes; betting everything on one throw, if unfortunately unsuccessful, there is no possibility of gambling again. Things proceed in circuits and twists, not in straight lines; warfare is the same—only formalists fail to comprehend this principle.

(110) I believe that even during the strategic counteroffensive stage, the principle of decisive engagements remains applicable. Although at that time the enemy is in a position of disadvantage and we are in a position of advantage, *the principle of "conducting favorable decisive engagements and avoiding unfavorable decisive engagements" still applies*, right up to reaching the Yalu River. In this way, we can always stand in a position of initiative; all enemy "challenges" and others' "provocations" should be consigned to the shelf, disregarded, and not allowed to move us in the slightest. Anti-Japanese generals must possess such firmness to be considered brave and wise generals. Those who "jump at the slightest provocation" are not worthy of discussion on this point. During the first stage, we occupy a certain degree of strategic passivity, yet on all campaigns we should also be active; thereafter, on any stage, we should be active. *We are proponents of protracted war and ultimate victory, not proponents of betting everything on one throw like gamblers.*

D. The Soldiers and the People as the Foundation of Victory (§§111–118)

(111) Japanese imperialism, confronted by revolutionary China, will by no means relax its offensive and repression; its imperialist essence dictates this point. If China does not resist, Japan can occupy China without expending a single bullet; the loss of the three northeastern provinces is a precedent. If China resists, Japan will press upon this resistance until its pressure cannot exceed China's resistance before ceasing—this is an inevitable law. The ambitions of Japan's landlord bourgeoisie are very great; in order to attack the Nanyang archipelago to the south and Siberia to the north, they adopt the policy of breakthrough in the center, attacking China first. Those who believe that Japan will stop after occupying North China and the Jiang-Zhe region completely fail to see that Japanese imperialism, having developed to a new stage and approached the brink of death, is no longer the same as historical Japan. When we say that Japan's troop deployment and offensive points have certain limitations, we mean: on Japan's side, based on its strength, in order to conduct offensives in other directions and defend against enemies on another front, it can only deploy a certain degree of force against China, up to the limit of its capability; on China's side, this also manifests our progress and tenacious resistance—we cannot conceive of only Japan launching fierce attacks while China lacks necessary resistance. Japan cannot occupy all of China; however, in all regions within its capability, it will spare no effort in repressing Chinese resistance, and will not cease this repression until Japan's internal and external conditions cause Japanese imperialism to enter a direct crisis leading to its grave. Japan's domestic politics has only two outlets: either the entire ruling class rapidly collapses, power is transferred to the people, and the war thus ends—but this is temporarily impossible; or the landlord bourgeoisie becomes increasingly fascist, sustaining the war until its own day of collapse—Japan is precisely taking this road. There is no third path. Those who hope that a centrist faction of the Japanese bourgeoisie will emerge to stop the war are merely indulging in fantasy. The centrist faction of Japan's bourgeoisie has long been the captive of the landlord and financial oligarchy; this is the actuality of Japanese politics over many years. After Japan fights China, if China's War of Resistance has not yet inflicted a fatal blow upon Japan, and Japan still possesses sufficient strength, it will certainly attack either Nanyang or Siberia, or even both. Once the European war begins, it will certainly take this action; the calculations of Japan's rulers are very ambitious. Of course,

the possibility exists: due to the strength of the Soviet Union, and due to Japan's great weakening in the war against China, Japan may have to abandon its original plan to attack Siberia and adopt a fundamentally defensive posture toward it. However, when such a situation arises, this will not mean a relaxation of Japan's offensive against China, but rather an intensification, because at that point it will have only one path left: devouring the weak. At that time, the tasks of China's persisting in resistance, persisting in the united front, and persisting in protracted war will appear even more grave, and not the slightest relaxation can be permitted.

(112) Under these circumstances, the principal conditions for China to defeat Japan are *national unity and progress in all respects tenfold or hundredfold beyond the past*. China has entered a progressive era and has already achieved great unity, but the current degree is still very insufficient. Japan occupies such extensive territory partly due to Japan's strength, partly due to China's weakness; and this weakness is entirely the result of the accumulation of various historical errors over the past century, especially the past decade, which has restricted China's progressive factors to today's state. To now defeat such a strong enemy is impossible without prolonged and extensive effort. There are many matters requiring effort; here I speak only of the two most fundamental aspects: *the progress of the army and the progress of the people*.

(113) Reform of the military system cannot be separated from modernization; enhancing technical conditions is indispensable—without this point, it is impossible to drive the enemy beyond the Yalu River. The employment of the army requires progressive, flexible strategy and tactics; without this point, victory is also impossible. However, *the foundation of the army lies in the soldiers*; without progressive political spirit infused into the army, without progressive political work to execute this infusion, it is impossible to achieve genuine unity between officers and soldiers, impossible to inspire the maximum anti-Japanese enthusiasm among officers and men, and impossible for all techniques and tactics to have the best foundation for exerting their due effectiveness. We say that although Japan's technical conditions are superior, it will ultimately fail; besides inflicting annihilative and attritive strikes upon it, its morale will inevitably waver under our strikes, and the combination of weapons and personnel will become unstable. We are the opposite: the political purpose of the War of Resistance Against Japan is unity between officers and

soldiers. On this foundation lies the basis for all political work in anti-Japanese armies. The army should implement a certain degree of democratization, principally abolishing the feudalistic system of beating and scolding and ensuring that officers and soldiers share weal and woe. In this way, the purpose of unity between officers and soldiers is achieved; the army gains immense combat effectiveness; and there is no fear that the long and cruel war cannot be sustained.

(114) *The deepest and most profound source of war's mighty power resides in the populace.* Japan dares to bully us principally because of the disorganized state of the Chinese populace. Overcoming this shortcoming places the Japanese aggressors before our hundreds of millions of risen people; it is like a wild bull charging into a ring of fire—we need only give a shout to startle it greatly, and this wild bull will inevitably be burned to death. On our side, the army must have a continuous supply of reinforcements; the current indiscriminate practices below of "catching soldiers" and "buying soldiers" [42] must be urgently prohibited and replaced by extensive and enthusiastic political mobilization—in this way, recruiting several million soldiers becomes easy. Anti-Japanese financial resources are extremely difficult; but with the populace mobilized, even financial matters pose no problem—how could a country with such vast territory and populous inhabitants suffer from financial exhaustion? The army must merge with the populace, so that the army is regarded by the populace as its own army; such an army would be invincible under heaven, and a mere Japanese imperialism would not be worth fighting.

(115) Many people, when unable to manage relationships between officers and soldiers or between army and people well, think the method is incorrect. I always tell them it is fundamentally a question of attitude (or fundamental purpose); this attitude is *respect for soldiers and respect for the people*. From this attitude proceed various policies, methods, and forms. Departing from this attitude, policies, methods, and forms will certainly be wrong, and relationships between officers and soldiers or between army and people will certainly not be managed well. The three major principles of army political work are: first, unity between officers and soldiers; second, unity between army and people; third, disintegration of enemy forces. For these principles to be effectively implemented, they must all proceed from the fundamental attitude of respecting soldiers, respecting the people, and respecting the personality of enemy prisoners who have laid down their arms. Those who think this is not a question of fundamental attitude but of technique are truly mistaken and should correct this view.

(116) At this time, when defending Wuhan and other locations has become an urgent task, mobilizing the entire army and entire populace to support the war is an extremely grave task. The task of defending Wuhan and other locations must undoubtedly be seriously proposed and executed. However, whether it can be definitively defended without loss is not decided by subjective wishes but by concrete conditions. Politically mobilizing the entire army and entire populace to struggle is one of the most important concrete conditions. If we do not strive to secure all necessary conditions—even if one necessary condition is lacking—we will inevitably repeat the mistake of losing Nanjing and other locations. Where is China's Madrid? It lies wherever the conditions of Madrid exist. In the past, there was never a Madrid; in the future, we should strive to win several—but it all depends on conditions. The most basic condition among conditions is *extensive political mobilization of the entire army and entire populace.*

(117) In all work, the general policy of the anti-Japanese national united front should be upheld. Because only this policy can persist in resistance, persist in protracted war, universally and profoundly improve relationships between officers and soldiers and between army and people, mobilize the entire enthusiasm of the entire army and entire populace, fight to defend all unlost regions and recover all lost regions, and strive for ultimate victory.

(118) This question of politically mobilizing the army and people is truly extremely important. The reason we speak of this point repeatedly and at length is that without this point, there is no victory. Although the absence of many other necessary things also means no victory, this is the most basic condition for victory. The anti-Japanese national united front is the united front of the entire army and entire populace; it is by no means merely the united front of the party organs and party members of several political parties; mobilizing the entire army and entire populace to participate in the united front is the fundamental purpose of initiating the anti-Japanese national united front.

E. Conclusion (§§119–120)

(119) What is the conclusion? The conclusion is:

"Under what conditions can China defeat and eliminate the power of Japanese imperialism? Three conditions are required: First, the completion of China's anti-Japanese united front; second, the completion of an international anti-Japanese united front; third, the rise of revolutionary movements among the Japanese people and the peoples of Japan's colonies. From the standpoint of the Chinese people, the great unity of the Chinese people is the primary condition among these three."

"How long will this war last? That depends on the strength of China's anti-Japanese united front and on many other decisive factors concerning China and Japan."

"If these conditions cannot be realized quickly, the war will be prolonged. Yet the outcome remains the same: Japan will inevitably be defeated, and China will inevitably triumph. The sacrifice, however, will be great, and a very arduous period must be endured."

"Our strategic guideline should be to employ our main forces to operate along extensive, fluid fronts. For the Chinese military to prevail, it must conduct highly mobile warfare across vast battlefields."

"In addition to deploying trained troops for mobile warfare, we must organize numerous guerrilla units among the peasantry."

"In the course of the war... the equipment of Chinese forces can be gradually strengthened. Consequently, China will be able to engage in positional warfare during the later stages of the conflict and launch positional attacks against Japanese-occupied territories. Thus, under the prolonged attrition of China's resistance, Japan's economy will approach collapse; amid countless military engagements, its morale will inevitably deteriorate. On China's side, the latent potential for resistance will surge ever higher day by day, as great numbers of revolutionary masses continuously pour to the front to fight for freedom. All these factors, combined with others, will enable us to launch a final, decisive assault against Japan's fortified positions and base areas in occupied China, and to expel the Japanese aggressors from our country."

"China's political situation has thereby entered a new stage... The central task of this stage is: to mobilize all forces to secure victory in the War of Resistance."

"The central key to achieving victory in the War of Resistance lies in developing the already-commenced resistance into a comprehensive, nationwide, people's war. Only such a comprehensive, nationwide, people's war can secure the ultimate victory of the resistance."

"Owing to serious weaknesses that still exist in the current resistance, many adverse developments may occur in the course of the war going forward: setbacks, retreats, internal fragmentation, defections, and temporary or partial compromises. Therefore, it must be recognized that this resistance is an arduous, protracted war. Yet we believe that the resistance already launched will, through the efforts of our Party and the entire Chinese people, overcome all obstacles and continue to advance and develop."

(August 1937, "Decision of the Central Committee of the Communist Party of China on the Current Situation and the Tasks of the Party")

These are the conclusions. Proponents of the theory of national subjugation regard the enemy as divine and ourselves as insignificant; proponents of the theory of quick victory regard the enemy as insignificant and ourselves as divine—both are erroneous. Our view is opposite: *the War of Resistance Against Japan is a protracted war, and ultimate victory belongs to China—this is our conclusion.*

(120) My address ends here. The great War of Resistance Against Japan is unfolding; many hope to summarize experience in order to strive for complete victory. What I have said represents only the general elements of ten months of experience; it may also be considered a summary. This question deserves the broad attention and discussion of all; what I have said is merely a general outline. I hope you will study, discuss, correct, and supplement it.

Glossary of Key Terms

Chinese Term	Formal English Translation	Strategic Significance
持久战	Protracted War	Core strategic concept: time as weapon against superior force
亡国论	Theory of National Subjugation / Defeatism	Primary ideological error refuted
速胜论	Theory of Quick Victory	Secondary ideological error refuted
半殖民地半封建	Semi-colonial and Semi-feudal	Defines China's social formation and material constraints
退步性 / 进步性	Retrogressive / Progressive Character	Historical-materialist criterion for war's justice
运动战 / 游击战 / 阵地战	Mobile / Guerrilla / Positional Warfare	Forms of warfare shifting by strategic stage
战略防御 / 相持 / 反攻	Strategic Defense / Stalemate / Counteroffensive	Three canonical stages of protracted war
内线 / 外线	Interior Lines / Exterior Lines	Spatial positioning relative to enemy forces
保存自己，消灭敌人	Preserve Oneself, Eliminate the Enemy	Fundamental purpose of all military action
主动性 / 灵活性 / 计划性	Initiative / Flexibility / Planning	Command trinity for operational success
兵民是胜利之本	Soldiers and People as the Foundation of Victory	Core political-military principle of people's war
抗日民族统一战线	Anti-Japanese National United Front	Overarching political strategy for victory

Footnote Reference Guide

(Markers preserved in translation)

Marker Reference Summary

[1]	Lugou Bridge Incident (July 7, 1937): start of full-scale Sino-Japanese War
[2]	Abyssinia (Ethiopia): reference to 1935–36 Italian invasion
[3]	"Further resistance will surely lead to perdition": defeatist slogan
[4]	"Mechanical viewpoint": critique of rigid strategic thinking
[5]	Reliance on foreign aid: critique of passive internationalism
[6]	Taierzhuang Victory (March–April 1938): major Chinese tactical success
[7]	Xuzhou Campaign: broader operational context of Taierzhuang
[8]	"Final struggle" rhetoric: example of impetuous optimism
[9]	Trotskyites: reference to internal political struggles within leftist movements
[10]–[42]	Historical events, persons, and concepts (full annotations available)

Citing this translation:

Mao, Zedong. *On Protracted War* [论持久战]. Speech delivered before the Yan'an Association for the Study of the War of Resistance Against Japan, May 26–June 3, 1938. Formal English translation prepared for *Institute for Integrated Systems Thinking*, 2026. Based on authoritative Chinese text from *Selected Works of Mao Zedong*, Vol. II. Footnote markers correspond to original scholarly apparatus.

Translators Final Note:

This translation standout for its adherence to the tone of the original text which was delivered as speech, without sacrificing the need for the formal tone to march its conceptual roots. This work is foundational text of revolutionary military theory, demonstrating how a materially weaker force can defeat a stronger adversary through political mobilization, strategic patience, and the dialectical transformation of disadvantages into advantages over time.

Original Text (Chinese)

论持久战

（一九三八年五月）

* 这是毛泽东一九三八年五月二十六日至六月三日在延安抗日战争研究会的讲演。

问题的提起

（一）伟大抗日战争的一周年纪念，七月七日，快要到了。全民族的力量团结起来，坚持抗战，坚持统一战线，同敌人作英勇的战争，快一年了。这个战争，在东方历史上是空前的，在世界历史上也将是伟大的，全世界人民都关心这个战争。身受战争灾难、为着自己民族的生存而奋斗的每一个中国人，无日不在渴望战争的胜利。然而战争的过程究竟会要怎么样？能胜利还是不能胜利？能速胜还是不能速胜？很多人都说持久战，但是为什么是持久战？怎样进行持久战？很多人都说最后胜利，但是为什么会有最后胜利？怎样争取最后胜利？这些问题，不是每个人都解决了的，甚至是大多数人至今没有解决的。于是失败主义的亡国论者跑出来向人们说：中国会亡，最后胜利不是中国的。某些性急的朋友们也跑出来向人们说：中国很快就能战胜，无需乎费大气力。这些议论究竟对不对呢？我们一向都说：这些议论是不对的。可是我们说的，还没有为大多数人所了解。一半因为我们的宣传解释工作还不够，一半也因为客观事变的发展还没有完全暴露其固有的性质，还没有将其面貌鲜明地摆在人们之前，使人们无从看出其整个的趋势和前途，因而无从决定自己的整套的方针和做法。现在好了，抗战十个月的经验，尽够击破毫无根据的亡国论，也尽够说服急性朋友们的速胜论了。在这种情形下，很多人要求做个总结性的解释。尤其是对持久战，有亡国论和速胜论的反对意见，也有空洞无物的了解。"卢沟桥事变[1]以来，四万万人一齐努力，最后胜利是中国的。"这样一种公式，在广大的人们中流行着。这个公式是对的，但有加以充实的必要。抗日战争和统一战线之所以能够坚持，是由于许多的因素：全国党派，从共产党到国民党；全国人民，从工人农民到资产阶级；全国军队，从主力军到游击队；国际方面，从社会主义国家到各国爱好正义的人民；敌国方面，从某些国内反战的人民到前线反战的兵士。总而言之，所有这些因素，在我们的抗战中都尽了他们各种程度的努力。每一个有良心的人，都应向他们表示敬意。我们共产党人，同其它抗战党派和全国人民一道，唯一的方向，是努力团结一切力量，战胜万恶的日寇。今年七月一日，是中国共产党建立的十七周年纪念日。为了使每个共产党员在抗日战争中能够尽其更好和更大的努力，也有着重地研究持久战的必要。因此，我的讲演就来研究持久战。和持久战这个题目有关的问题，我都准备说到；但是不能一切都说到，因为一切的东西，不是在一个讲演中完全说得了的。

（二）抗战十个月以来，一切经验都证明下述两种观点的不对：一种是中国必亡论，一种是中国速胜论。前者产生妥协倾向，后者产生轻敌倾向。他们看问题的方法都是主观的和片面的，一句话，非科学的。

（三）抗战以前，存在着许多亡国论的议论。例如说："中国武器不如人，战必败。""如果抗战，必会作阿比西尼亚[2]。"抗战以后，公开的亡国论没有了，但暗地是有的，而且很多。例如妥协的空气时起时伏，主张妥协者的根据就是"再战必亡"[3]。有个学生从湖南写信来说："在乡下一切都感到困难。单独一个人作宣传工作，只好随时随地找人谈话。对象都不是无知无识的愚民，他们多少也懂得一点，他们对我的谈话很有兴趣。可是碰了我那几位亲戚，他们总说：'中国打不胜，会亡。'讨厌极了。好在他们还不去宣传，不然真糟。农民对他们的信仰当然要大些啊！"这类中国必亡论者，是妥协倾向的社会基础。这类人中国各地都有，因此，抗日阵线中随时可能发生的妥协问题，恐怕终战争之局也不会消灭的。当此徐州失守武汉紧张的时候，给这种亡国论痛驳一驳，我想不是无益的。

（四）抗战十个月以来，各种表现急性病的意见也发生了。例如在抗战初起时，许多人有一种毫无根据的乐观倾向，他们把日本估计过低，甚至以为日本不能打到山西。有些人轻视抗日战争中游击战争的战略地位，他们对于"在全体上，运动战是主要的，游击战是辅助的；在部分上，游击战是主要的，运动战是辅助的"这个提法，表示怀疑。他们不赞成八路军这样的战略方针："基本的是游击战，但不放松有利条件下的运动战。"认为这是"机械的"观点[4]。上海战争时，有些人说："只要打三个月，国际局势一定变化，苏联一定出兵，战争就可解决。"把抗战的前途主要地寄托在外国援助上面[5]。台儿庄胜利[6]之后，有些人主张徐州战役[7]应是"准决战"，说过去的持久战方针应该改变。说什么"这一战，就是敌人的最后挣扎"，"我们胜了，日阀就在精神上失了立场，只有静候末日审判"[8]。平型关一个胜仗，冲昏了一些人的头脑；台儿庄再一个胜仗，冲昏了更多的人的头脑。于是敌人是否进攻武汉，成为疑问了。许多人以为："不一定"；许多人以为："断不会"。这样的疑问可以牵涉到一切重大的问题。例如说：抗日力量是否够了呢？回答可以是肯定的，因为现在的力量已使敌人不能再进攻，还要增加力量干什么呢？例如说：巩固和扩大抗日民族统一战线的口号是否依然正确呢？回答可以是否定的，因为统一战线的现时状态已够打退敌人，还要什么巩固和扩大呢？例如说：国际外交和国际宣传工作是否还应该加紧呢？回答也可以是否定的。例如说：改革军队制度，改革政治制度，发展民众运动，厉行国防教育，镇压汉奸托派[9]，发展军事工业，改良人民生活，是否应该认真去做呢？例如说：保卫武汉、保卫广州、保卫西北和猛烈发展敌后游击战争的口号，是否依然正确呢？回答都可以是否定的。甚至某些人在战争形势稍为好转的时候，就准备在国共两党之间加紧磨擦一下，把对外的眼光转到对内。这种情况，差不多每一个较大的胜仗之后，或敌人进攻暂时停顿之时，都要发生。所有上述一切，我们叫它做政治上军事上的近视眼。这些话，讲起来好像有道理，实际上是毫无根据、似是而非的空谈。扫除这些空谈，对于进行胜利的抗日战争，应该是有好处的。

（五）于是问题是：中国会亡吗？答复：不会亡，最后胜利是中国的。中国能够速胜吗？答复：不能速胜，抗日战争是持久战。

（六）这些问题的主要论点，还在两年之前我们就一般地指出了。还在一九三六年七月十六日，即在西安事变前五个月，卢沟桥事变前十二个月，我同美国记者斯诺先生的谈话中，就已经一般地估计了中日战争的形势，并提出了争取胜利的各种方针。为备忘计，不妨抄录几段如下：

> 问：在什么条件下，中国能战胜并消灭日本帝国主义的实力呢？

> 答：要有三个条件：第一是中国抗日统一战线的完成；第二是国际抗日统一战线的完成；第三是日本国内人民和日本殖民地人民的革命运动的兴起。就中国人民的立场来说，三个条件中，中国人民的大联合是主要的。

> 问：你想，这个战争要延长多久呢？

> 答：要看中国抗日统一战线的实力和中日两国其它许多决定的因素如何而定。即是说，除了主要地看中国自己的力量之外，国际间所给中国的援助和日本国内革命的援助也很有关系。如果中国抗日统一战

线有力地发展起来，横的方面和纵的方面都有效地组织起来，如果认清日本帝国主义威胁他们自己利益的各国政府和各国人民能给中国以必要的援助，如果日本的革命起来得快，则这次战争将迅速结束，中国将迅速胜利。如果这些条件不能很快实现，战争就要延长。但结果还是一样，日本必败，中国必胜。只是牺牲会大，要经过一个很痛苦的时期。

问：从政治上和军事上来看，你以为这个战争的前途会要如何发展？

答：日本的大陆政策已经确定了，那些以为同日本妥协，再牺牲一些中国的领土主权就能够停止日本进攻的人们，他们的想法只是一种幻想。我们确切地知道，就是扬子江下游和南方各港口，都已经包括在日本帝国主义的大陆政策之内。并且日本还想占领菲律宾、暹罗、越南、马来半岛和荷属东印度，把外国和中国切开，独占西南太平洋。这又是日本的海洋政策。在这样的时期，中国无疑地要处于极端困难的地位。可是大多数中国人相信，这种困难是能够克服的；只有各大商埠的富人是失败论者，因为他们害怕损失财产。有许多人想，一旦中国海岸被日本封锁，中国就不能继续作战。这是废话。为反驳他们，我们不妨举出红军的战争史。在抗日战争中，中国所占的优势，比内战时红军的地位强得多。中国是一个庞大的国家，就是日本能占领中国一万万至二万万人口的区域，我们离战败还很远呢。我们仍然有很大的力量同日本作战，而日本在整个战争中须得时时在其后方作防御战。中国经济的不统一、不平衡，对于抗日战争反为有利。例如将上海和中国其它地方割断，对于中国的损害，绝没有将纽约和美国其它地方割断对于美国的损害那样严重。日本就是把中国沿海封锁，中国的西北、西南和西部，它是无法封锁的。所以问题的中心点还是中国全体人民团结起来，树立举国一致的抗日阵线。这是我们早就提出了的。

问：假如战争拖得很长，日本没有完全战败，共产党能否同意讲和，并承认日本统治东北？

答：不能。中国共产党和全国人民一样，不容许日本保留中国的寸土。

问：照你的意见，这次解放战争，主要的战略方针是什么？

答：我们的战略方针，应该是使用我们的主力在很长的变动不定的战线上作战。中国军队要胜利，必须在广阔的战场上进行高度的运动战，迅速地前进和迅速地后退，迅速地集中和迅速地分散。这就是大规模的运动战，而不是深沟高垒、层层设防、专靠防御工事的阵地战。这并不是说要放弃一切重要的军事地点，对于这些地点，只要有利，就应配置阵地战。但是转换全局的战略方针，必然要是运动战。阵地战虽也必需，但是属于辅助性质的第二种的方针。在地理上，战场这样广大，我们作最有效的运动战，是可能的。日军遇到我军的猛烈活动，必得谨慎。他们的战争机构很笨重，行动很慢，效力有限。如果我们集中兵力在一个狭小的阵地上作消耗战的抵抗，将使我军失掉地理上和经济组织上的有利条件，犯阿比西尼亚的错误。战争的前期，我们要避免一切大的决战，要先用运动战逐渐地破坏敌人军队的精神和战斗力。

除了调动有训练的军队进行运动战之外，还要在农民中组织很多的游击队。须知东三省的抗日义勇军，仅仅是表示了全国农民所能动员抗战的潜伏力量的一小部分。中国农民有很大的潜伏力，只要组织和指挥得当，能使日本军队一天忙碌二十四小时，使之疲于奔命。必须记住这个战争是在中国打的，这就是说，日军要完全被敌对的中国人所包围；日军要被迫运来他们所需的军用品，而且要自己看守；他们要用重兵去保护交通线，时时谨防袭击；另外，还要有一大部力量驻扎满洲和日本内地。

在战争的过程中，中国能俘虏许多的日本兵，夺取许多的武器弹药来武装自己；同时，争取外国的援助，使中国军队的装备逐渐加强起来。因此，中国能够在战争的后期从事阵地战，对于日本的占领地进行阵地的攻击。这样，日本在中国抗战的长期消耗下，它的经济行将崩溃；在无数战争的消磨中，它的士气行将颓靡。中国方面，则抗战的潜伏力一天一天地奔腾高涨，大批的革命民众不断地倾注到前线去，为自由而战争。所有这些因素和其它的因素配合起来，就使我们能够对日本占领地的堡垒和根据地，作最后的致命的攻击，驱逐日本侵略军出中国。（斯诺：《西北印象记》）

抗战十个月的经验，证明上述论点的正确，以后也还将继续证明它。

（七）还在卢沟桥事变发生后一个多月，即一九三七年八月二十五日，中国共产党中央就在它的《关于目前形势与党的任务的决定》中，清楚地指出：

> 卢沟桥的挑战和平津的占领，不过是日寇大举进攻中国本部的开始。日寇已经开始了全国的战时动员。他们的所谓"不求扩大"的宣传，不过是掩护其进攻的烟幕弹。

> 七月七日卢沟桥的抗战，已经成了中国全国性抗战的起点。

> 中国的政治形势从此开始了一个新阶段，这就是实行抗战的阶段。抗战的准备阶段已经过去了。这一阶段的最中心的任务是：动员一切力量争取抗战的胜利。

> 争取抗战胜利的中心关键，在使已经发动的抗战发展为全面的全民族的抗战。只有这种全面的全民族的抗战，才能使抗战得到最后的胜利。

> 由于当前的抗战还存在着严重的弱点，所以在今后的抗战过程中，可能发生许多挫败、退却，内部的分化、叛变，暂时和局部的妥协等不利的情况。因此，应该看到这一抗战是艰苦的持久战。但我们相信，已经发动的抗战，必将因为我党和全国人民的努力，冲破一切障碍物而继续地前进和发展。

抗战十个月的经验，同样证明了上述论点的正确，以后也还将继续证明它。

（八）战争问题中的唯心论和机械论的倾向，是一切错误观点的认识论上的根源。他们看问题的方法是主观的和片面的。或者是毫无根据地纯主观地说一顿；或者是只根据问题的一侧面、一时候的表现，也同样主观地把它夸大起来，当作全体看。但是人们的错误观点可分为两类：一类是根本的错误，带一贯性，这是难于纠正的；另一类是偶然的错误，带暂时性，这是易于纠正的。但既同为错误，就都有纠正的必要。因此，反对战争问题中的唯心论和机械论的倾向，采用客观的观点和全面的观点去考察战争，才能使战争问题得出正确的结论。

问题的根据

（九）抗日战争为什么是持久战？最后胜利为什么是中国的呢？根据在什么地方呢？

中日战争不是任何别的战争，乃是半殖民地半封建的中国和帝国主义的日本之间在二十世纪三十年代进行的一个决死的战争。全部问题的根据就在这里。分别地说来，战争的双方有如下互相反对的许多特点。

（一〇）日本方面：第一，它是一个强的帝国主义国家，它的军力、经济力和政治组织力在东方是一等的，在世界也是五六个著名帝国主义国家中的一个。这是日本侵略战争的基本条件，战争的不可避免和中国的不能速胜，就建立在这个日本国家的帝国主义制度及其强的军力、经济力和政治组织力上面。然而第二，由于日本社会经济的帝国主义性，就产生了日本战争的帝国主义性，它的战争是退步的和野蛮的。时至二十世纪三十年代的日本帝国主义，由于内外矛盾，不但使得它不得不举行空前大规模的冒险战争，而且使得它临到最后崩溃的前夜。从社会行程说来，日本已不是兴旺的国家，战争不能达到日本统治阶级所期求的兴旺，而将达到它所期求的反面——日本帝国主义的死亡。这就是所谓日本战争的退步性。跟着这个退步性，加上日本又是一个带军事封建性的帝国主义这一特点，就产生了它的战争的特殊的野蛮性。这样就要最大地激起它国内的阶级对立、日本民族和中国民族的对立、日本和世界大多数国家的对立。日本战争的退步性和野蛮性是日本战争必然失败的主要根据。还不止此，第三，日本战争虽是在其强的军力、经济力和政治组织力的基础之上进行的，但同时又是在其先天不足的基础之上进行的。日本的军力、经济力和政治组织力虽强，但这些力量之量的方面不足。日本国度比较地小，其人力、军力、财力、物力均感缺乏，经不起长期的战争。日本统治者想从战争中解决这个困难问题，但同样，将达到其所期求的反面，这就是说，它为解决这个困难问题而发动战争，结果将因战争而增加困难，战争将连它原有的东西也消耗掉。最后，第四，日本虽能得到国际法西斯国家的援助，但同时，却又不能不遇到一个超过其国际援助力量的国际反对力量。这后一种力量将逐渐地增长，终究不但将把前者的援助力量抵消，并将施其压力于日本自身。这是失道寡助的规律，是从日本战争的本性产生出来的。总起来说，日本的长处是其战争力量之强，而其短处则在其战争本质的退步性、野蛮性，在其人力、物力之不足，在其国际形势之寡助。这些就是日本方面的特点。

（一一）中国方面：第一，我们是一个半殖民地半封建的国家。从鸦片战争[10]，太平天国[11]，戊戌维新[12]，辛亥革命[13]，直至北伐战争，一切为解除半殖民地半封建地位的革命的或改良的运动，都遭到了严重的挫折，因此依然保留下这个半殖民地半封建的地位。我们依然是一个弱国，我们在军力、经济力和政治组织力各方面都显得不如敌人。战争之不可避免和中国之不能速胜，又在这个方面有其基础。然而第二，中国近百年的解放运动积累到了今日，已经不同于任何历史时期。各种内外反对力量虽给了解放运动以严重挫折，同时却锻炼了中国人民。今日中国的军事、经济、政治、文化虽不如日本之强，但在中国自己比较起来，却有了比任何一个历史时期更为进步的因素。中国共产党及其领导下的军队，就是这种进步因素的代表。中国今天的解放战争，就是在这种进步的基础上得到了持久战和最后胜利的可能性。中国是如日方升的国家，这同日本帝国主义的没落状态恰是相反的对照。中国的战争是进步的，从这种进步性，就产生了中国战争的正义性。因为这个战争是正义的，就能唤起全国的团结，激起敌国人民的同情，争取世界多数国家的援助。第三，中国又是一个很大的国家，地大、物博、人多、兵多，能够支持长期的战争，这同日本又是一个相反的对比。最后，第四，由于中国战争的进步性、正义性而产生出来的国际广大援助，同日本的失道寡助又恰恰相反。总起来说，中国的短处是战争力量之弱，而其长处则在其战争本质的进步性和正义性，在其是一个大国家，在其国际形势之多助。这些都是中国的特点。

（一二）这样看来，日本的军力、经济力和政治组织力是强的，但其战争是退步的、野蛮的，人力、物力又不充足，国际形势又处于不利。中国反是，军力、经济力和政治组织力是比较地弱的，然而正处于进步的时代，其战争是进步的和正义的，又有大国这个条件足以支持持久战，世界的多数国家是会要援助中国的。——这些，就是中日战争互相矛盾着的基本特点。这些特点，规定了和规定着双方一切政治上的政策和军事上的战略战术，规定了和规定着战争的持久性和最后胜利属于中国而不属于日本。战争就是这些特点的比赛。这些特点在战争过程中将各依其本性发生变化，一切东西就都从这里发生出来。这些特点是事实上存在的，不是虚造骗人的；是战争的全部基本要素，不是残缺不全的片段；是贯彻于双方一切大小问题和一切作战阶段之中的，不是可有可无的。观察中日战争如果忘记了这些特点，那就必然要弄错；即使某些意见一时有人相信，似乎不错，但战争的经过必将证明它们是错的。我们现在就根据这些特点来说明我们所要说的一切问题。

驳亡国论

（一三）亡国论者看到敌我强弱对比一个因素，从前就说"抗战必亡"，现在又说"再战必亡"。如果我们仅仅说，敌人虽强，但是小国，中国虽弱，但是大国，是不足以折服他们的。他们可以搬出元朝灭宋、清朝灭明的历史证据，证明小而强的国家能够灭亡大而弱的国家，而且是落后的灭亡进步的。如果我们说，这是古代，不足为据，他们又可以搬出英灭印度的事实，证明小而强的资本主义国家能够灭亡大而弱的落后国家。所以还须提出其它的根据，才能把一切亡国论者的口封住，使他们心服，而使一切从事宣传工作的人们得到充足的论据去说服还不明白和还不坚定的人们，巩固其抗战的信心。

（一四）这应该提出的根据是什么呢？就是时代的特点。这个特点的具体反映是日本的退步和寡助，中国的进步和多助。

（一五）我们的战争不是任何别的战争，乃是中日两国在二十世纪三十年代进行的战争。在我们的敌人方面，首先，它是快要死亡的帝国主义，它已处于退步时代，不但和英灭印度时期英国还处于资本主义的进步时代不相同，就是和二十年前第一次世界大战时的日本也不相同。此次战争发动于世界帝国主义首先是法西斯国家大崩溃的前夜，敌人也正是为了这一点才举行这个带最后挣扎性的冒险战争。所以，战争的结果，灭亡的不会是中国而是日本帝国主义的统治集团，这是无可逃避的必然性。再则，当日本举行战争的时候，正是世界各国或者已经遭遇战争或者快要遭遇战争的时候，大家都正在或准备着为反抗野蛮侵略而战，中国这个国家又是同世界多数国家和多数人民利害相关的，这就是日本已经引起并还要加深地引起世界多数国家和多数人民的反对的根源。

（一六）中国方面呢？它已经不能和别的任何历史时期相比较。半殖民地和半封建社会是它的特点，所以被称为弱国。但是在同时，它又处于历史上进步的时代，这就是足以战胜日本的主要根据。所谓抗日战争是进步的，不是说普通一般的进步，不是说阿比西尼亚抗意战争的那种进步，也不是说太平天国或辛亥革命的那种进步，而是说今天中国的进步。今天中国的进步在什么地方呢？在于它已经不是完全的封建国家，已经有了资本主义，有了资产阶级和无产阶级，有了已经觉悟或正在觉悟的广大人民，有了共产党，有了政治上进步的军队即共产党领导的中国红军，有了数十年革命的传统经验，特别是中国共产党成立以来的十七年的经验。这些经验，教育了中国的人民，教育了中国的政党，今天恰好作了团结抗日的基础。如果说，在俄国，没有一九〇五年的经验就不会有一九一七年的胜利；那末，我们也可以说，如果没有十七年以来的经验，也将不会有抗日的胜利。这是国内的条件。

国际的条件，使得中国在战争中不是孤立的，这一点也是历史上空前的东西。历史上不论中国的战争也罢，印度的战争也罢，都是孤立的。惟独今天遇到世界上已经发生或正在发生的空前广大和空前深刻的人民运动及其对于中国的援助。俄国一九一七年的革命也遇到世界的援助，俄国的工人和农民因此胜利了，但那个援助的规模还没有今天广大，性质也没有今天深刻。今天的世界的人民运动，正在以空前的大规模和空前的深刻性发展着。苏联的存在，更是今天国际政治上十分重要的因素，它必然以极大的热忱援助中国，这一现象，是二十年前完全没有的。所有这些，造成了和造成着为中国最后胜利所不可缺少的重要的条件。大量的直接的援助，目前虽然还没有，尚有待于来日，但是中国有进步和大国的条件，能够延长战争的时间，促进并等候国际的援助。

（一七）加上日本是小国，地小、物少、人少、兵少，中国是大国，地大、物博、人多、兵多这一个条件，于是在强弱对比之外，就还有小国、退步、寡助和大国、进步、多助的对比，这就是中国决不会亡的根据。强弱对比虽然规定了日本能够在中国有一定时期和一定程度的横行，中国不可避免地要走一段艰难的路程，抗日战争是持久战而不是速决战；然而小国、退步、寡助和大国、进步、多助的对比，又规定了日本不能横行到底，必然要遭到最后的失败，中国决不会亡，必然要取得最后的胜利。

（一八）阿比西尼亚为什么灭亡了呢？第一，它不但是弱国，而且是小国。第二，它不如中国进步，它是一个古老的奴隶制到农奴制的国家，没有资本主义，没有资产阶级政党，更没有共产党，没有中国这样的军队，更没有如同八路军这样的军队。第三，它不能等候国际的援助，它的战争是孤立的。第四，这是主要的，抗意战争领导方面有错误。阿比西尼亚因此灭亡了。然而阿比西尼亚还有相当广大的游击战争存在，如能坚持下去，是可以在未来的世界变动中据以恢复其祖国的。

（一九）如果亡国论者搬出中国近代解放运动的失败史来证明"抗战必亡"和"再战必亡"的话，那我们的答复也是时代不同一句话。中国本身、日本内部、国际环境都和过去不相同。日本比过去更强了，中国的半殖民地和半封建地位依然未变，力量依然颇弱，这一点是严重的情形。日本暂时还能控制其国内的人民，也还能利用国际间的矛盾作为其侵华的工具，这些都是事实。然而在长期的战争过程中，必然要发生相反的变化。这一点现在还不是事实，但是将来必然要成为事实的。这一点，亡国论者就抛弃不顾了。中国呢？不但现在已有新的人、新的政党、新的军队和新的抗日政策，和十余年以前有很大的不同，而且这些都必然会向前发展。虽然历史上的解放运动屡次遭受挫折，使中国不能积蓄更大的力量用于今日的抗日战争——这是非常可痛惜的历史的教训，从今以后，再也不要自己摧残任何的革命力量了——然而就在既存的基础上，加上广大的努力，必能逐渐前进，加强抗战的力量。伟大的抗日民族统一战线，就是这种努力的总方向。国际援助一方面，眼前虽然还看不见大量的和直接的，但是国际局面根本已和过去两样，大量和直接的援助正在酝酿中。中国近代无数解放运动的失败都有其客观和主观的原因，都不能比拟今天的情况。在今天，虽然存在着许多困难条件，规定了抗日战争是艰难的战争，例如敌人之强，我们之弱，敌人的困难还刚在开始，我们的进步还很不够，如此等等，然而战胜敌人的有利条件是很多的，只须加上主观的努力，就能克服困难而争取胜利。这些有利条件，历史上没有一个时候可和今天比拟，这就是抗日战争必不会和历史上的解放运动同归失败的理由。

妥协还是抗战？腐败还是进步？

（二〇）亡国论之没有根据，俱如上述。但是另有许多人，并非亡国论者，他们是爱国志士，却对时局怀抱甚深的忧虑。他们的问题有两个：一是惧怕对日妥协，一是怀疑政治不能进步。这两个可忧虑的问题在广大的人们中间议论着，找不到解决的基点。我们现在就来研究这两个问题。

（二一）前头说过，妥协的问题是有其社会根源的，这个社会根源存在，妥协问题就不会不发生。但妥协是不会成功的。要证明这一点，仍不外向日本、中国、国际三方面找根据。第一是日本方面。还在抗战初起时，我们就估计有一种酝酿妥协空气的时机会要到来，那就是在敌人占领华北和江浙之后，可能出以劝降手段。后来果然来了这一手；但是危机随即过去，原因之一是敌人采取了普遍的野蛮政策，实行公开的掠夺。中国降了，任何人都要做亡国奴。敌人的这一掠夺的即灭亡中国的政策，分为物质的和精神的两方面，都是普遍地施之于中国人的；不但是对下层民众，而且是对上层成分，——当然对后者稍为客气些，但也只有程度之别，并无原则之分。大体上，敌人是将东三省的老办法移植于内地。在物质上，掠夺普通人民的衣食，使广大人民啼饥号寒；掠夺生产工具，使中国民族工业归于毁灭和奴役化。在精神上，摧残中国人民的民族意识。在太阳旗下，每个中国人只能当顺民，做牛马，不许有一丝一毫的中国气。敌人的这一野蛮政策，还要施之于更深的内地。他的胃口很旺，不愿停止战争。一九三八年一月十六日日本内阁宣言的方针[14]，至今坚决执行，也不能不执行，这就激怒了一切阶层的中国人。这是根据敌人战争的退步性野蛮性而来的，"在劫难逃"，于是形成了绝对的敌对。估计到某种时机，敌之劝降手段又将出现，某些亡国论者又将蠕蠕而动，而且难免勾结某些国际成分（英、美、法内部都有这种人，特别是英国的上层分子），狼狈为奸。但是大势所趋，是降不了的，日本战争的坚决性和特殊的野蛮性，规定了这个问题的一方面。

（二二）第二是中国方面。中国坚持抗战的因素有三个：其一，共产党，这是领导人民抗日的可靠力量。又其一，国民党，因其是依靠英美的，英美不叫它投降，它也就不会投降。又其一，别的党派，大多数是反对妥协、拥护抗战的。这三者互相团结，谁要妥协就是站在汉奸方面，人人得而诛之。一切不愿当汉奸的人，就不能不团结起来坚持抗战到底，妥协就实际上难于成功。

（二三）第三是国际方面。除日本的盟友和各资本主义国家的上层分子中的某些成分外，其余都不利于中国妥协而利于中国抗战。这一因素影响到中国的希望。今天全国人民有一种希望，认为国际力量必将逐渐增强地援助中国。这种希望不是空的；特别是苏联的存在，鼓舞了中国的抗战。空前强大的社会主义的苏联，它和中国是历来休戚相关的。苏联和一切资本主义国家的上层成分之唯利是图者根本相反，它是以援助一切弱小民族和革命战争为其职志的。中国战争之非孤立性，不但一般地建立在整个国际的援助上，而且特殊地建立在苏联的援助上。中苏两国是地理接近的，这一点加重了日本的危机，便利了中国的抗战。中日两国地理接近，加重了中国抗战的困难。然而中苏的地理接近，却是中国抗战的有利条件。

（二四）由此可作结论：妥协的危机是存在的，但是能够克服。因为敌人的政策即使可作某种程度的改变，但其根本改变是不可能的。中国内部有妥协的社会根源，但是反对妥协的占大多数。国际力量也有一部分赞助妥协，但是主要的力量赞助抗战。这三种因素结合起来，就能克服妥协危机，坚持抗战到底。

（二五）现在来答复第二个问题。国内政治的改进，是和抗战的坚持不能分离的。政治越改进，抗战越能坚持；抗战越坚持，政治就越能改进。但是基本上依赖于坚持抗战。国民党的各方面的不良现象是严重地存在着，这些不合理因素的历史积累，使得广大爱国志士发生很大的忧虑和烦闷。但是抗战的经验已经证明，十个月的中国人民的进步抵得上过去多少年的进步，并无使人悲观的根据。历史积累下来的腐败现象，虽然很严重地阻碍着人民抗战力量增长的速度，减少了战争的胜利，招致了战争的损失，但是中国、日本和世界的大局，不容许中国人民不进步。由于阻碍进步的因素即腐败现象之存在，这种进步是缓慢的。进步和进步的缓慢是目前时局的两个特点，后一个特点和战争的迫切要求很不相称，这就是使得爱国志士们大为发愁的地方。然而我们是在革命战争中，革命战争是一种抗毒素，它不但将排除敌人的毒焰，也将清洗自己的污浊。凡属正义的革命的战争，其力量是很大的，它能改造很多事物，或为改造事物开辟道路。中日战争将改造中日两国；只要中国坚持抗战和坚持统一战线，就一定能把旧日本化为新日本，把旧中国化为新中国，中日两国的人和物都将在这次战争中和战争后获得改造。我们把抗战和建国联系起来看，是正当的。说日本也能获得改造，是说日本统治者的侵略战争将走到失败，有引起日本人民革命之可能。日本人民革命胜利之日，就是日本改造之时。这和中国的抗战密切地联系着，这一个前途是应该看到的。

亡国论是不对的，速胜论也是不对的

（二六）我们已把强弱、大小、进步退步、多助寡助几个敌我之间矛盾着的基本特点，作了比较研究，批驳了亡国论，答复了为什么不易妥协和为什么政治可能进步的问题。亡国论者看重了强弱一个矛盾，把它夸大起来作为全部问题的论据，而忽略了其它的矛盾。他们只提强弱对比一点，是他们的片面性；他们将此片面的东西夸大起来看成全体，又是他们的主观性。所以在全体说来，他们是没有根据的，是错误的。那些并非亡国论者，也不是一贯的悲观主义者，仅为一时候和一局部的敌我强弱情况或国内腐败现象所迷惑，而一时地发生悲观心理的人们，我们也得向他们指出，他们的观点的来源也是片面性和主观性的倾向。但是他们的改正较容易，只要一提醒就会明白，因为他们是爱国志士，他们的错误是一时的。

（二七）然而速胜论者也是不对的。他们或则根本忘记了强弱这个矛盾，而单单记起了其它矛盾；或则对于中国的长处，夸大得离开了真实情况，变成另一种样子；或则拿一时一地的强弱现象代替了全体中的强弱现象，一叶障目，不见泰山，而自以为是。总之，他们没有勇气承认敌强我弱这件事实。他们常常抹杀这一点，因此抹杀了真理的一方面。他们又没有勇气承认自己长处之有限性，因而抹杀了真理的又一方面。由此犯出或大或小的错误来，这里也是主观性和片面性作怪。这些朋友们的心是好的，他们也是爱国志士。但是"先生之志则大矣"，先生的看法则不对，照了做去，一定碰壁。因为估计不符合真相，行动就无法达到目的；勉强行去，败军亡国，结果和失败主义者没有两样。所以也是要不得的。

（二八）我们是否否认亡国危险呢？不否认的。我们承认在中国面前摆着解放和亡国两个可能的前途，两者在猛烈地斗争中。我们的任务在于实现解放而避免亡国。实现解放的条件，基本的是中国的进步，同时，加上敌人的困难和世界的援助。我们和亡国论者不同，我们客观地而且全面地承认亡国和解放两个可能同时存在，着重指出解放的可能占优势及达到解放的条件，并为争取这些条件而努力。亡国论者则主观地和片面地只承认亡国一个可能性，否认解放的可能性，更不会指出解放的条件和为争取这些条件而努力。我们对于妥协倾向和腐败现象也是承认的，但是我们还看到其它倾向和其它现象，并指出二者之中后者对于前者将逐步地占优势，二者在猛烈地斗争着；并指出后者实现的条件，为克服妥协倾向和转变腐败现象而努力。因此，我们并不悲观，而悲观的人们则与此相反。

（二九）我们也不是不喜欢速胜，谁也赞成明天一个早上就把"鬼子"赶出去。但是我们指出，没有一定的条件，速胜只存在于头脑之中，客观上是不存在的，只是幻想和假道理。因此，我们客观地并全面地估计到一切敌我情况，指出只有战略的持久战才是争取最后胜利的唯一途径，而排斥毫无根据的速胜论。我们主张为着争取最后胜利所必要的一切条件而努力，条件多具备一分，早具备一日，胜利的把握就多一分，胜利的时间就早一日。我们认为只有这样才能缩短战争的过程，而排斥贪便宜尚空谈的速胜论。

为什么是持久战?

（三〇）现在我们来把持久战问题研究一下。"为什么是持久战"这一个问题，只有依据全部敌我对比的基本因素，才能得出正确的回答。例如单说敌人是帝国主义的强国，我们是半殖民地半封建的弱国，就有陷入亡国论的危险。因为单纯地以弱敌强，无论在理论上，在实际上，都不能产生持久的结果。单是大小或单是进步退步、多助寡助，也是一样。大并小、小并大的事都是常有的。进步的国家或事物，如果力量不强，常有被大而退步的国家或事物所灭亡者。多助寡助是重要因素，但是附随因素，依敌我本身的基本因素如何而定其作用的大小。因此，我们说抗日战争是持久战，是从全部敌我因素的相互关系产生的结论。敌强我弱，我有灭亡的危险。但敌尚有其它缺点，我尚有其它优点。敌之优点可因我之努力而使之削弱，其缺点亦可因我之努力而使之扩大。我方反是，我之优点可因我之努力而加强，缺点则因我之努力而克服。所以我能最后胜利，避免灭亡，敌则将最后失败，而不能避免整个帝国主义制度的崩溃。

（三一）既然敌之优点只有一个，余皆缺点，我之缺点只有一个，余皆优点，为什么不能得出平衡结果，反而造成了现时敌之优势我之劣势呢？很明显的，不能这样形式地看问题。事情是现时敌我强弱的程度悬殊太大，敌之缺点一时还没有也不能发展到足以减杀其强的因素之必要的程度，我之优点一时也没有且不能发展到足以补充其弱的因素之必要的程度，所以平衡不能出现，而出现的是不平衡。

（三二）敌强我弱，敌是优势而我是劣势，这种情况，虽因我之坚持抗战和坚持统一战线的努力而有所变化，但是还没有产生基本的变化。所以，在战争的一定阶段上，敌能得到一定程度的胜利，我则将遭到一定程度的失败。然而敌我都只限于这一定阶段内一定程度上的胜或败，不能超过而至于全胜或全败，这是什么缘故呢？因为一则敌强我弱之原来状况就是相对的，不是绝对的；二则由于我之坚持抗战和坚持统一战线的努力，更加造成这种相对的形势。拿原来状况来说，敌虽强，但敌之强已为其它不利的因素所减杀，不过此时还没有减杀到足以破坏敌之优势的必要的程度；我虽弱，但我之弱已为其它有利的因素所补充，不过此时还没有补充到足以改变我之劣势的必要的程度。于是形成敌是相对的强，我是相对的弱；敌是相对的优势，我是相对的劣势。双方的强弱优劣原来都不是绝对的，加以战争过程中我之坚持抗战和坚持统一战线的努力，更加变化了敌我原来强弱优劣的形势，因而敌我只限于一定阶段内的一定程度上的胜或败，造成了持久战的局面。

（三三）然而情况是继续变化的。战争过程中，只要我能运用正确的军事的和政治的策略，不犯原则的错误，竭尽最善的努力，敌之不利因素和我之有利因素均将随战争之延长而发展，必能继续改变着敌我强弱的原来程度，继续变化着敌我的优劣形势。到了新的一定阶段时，就将发生强弱程度上和优劣形势上的大变化，而达到敌败我胜的结果。

（三四）目前敌尚能勉强利用其强的因素，我之抗战尚未给他以基本的削弱。其人力、物力不足的因素尚不足以阻止其进攻，反之，尚足以维持其进攻到一定的程度。其足以加剧本国阶级对立和中国民族反抗的因素，即战争之退步性和野蛮性一因素，亦尚未造成足以根本妨碍其进攻的情况。敌人的国际孤立的因素也方在变化发展之中，还没有达到完全的孤立。许多表示助我的国家的军火资本家和战争原料资本家，尚在唯利是图地供给日本以大量的战争物资[15]，他们的政府[16]亦尚不愿和苏联一道用实际方法制裁日本。这一切，规定了我之抗战不能速胜，而只能是持久战。中国方面，弱的因素表现在军事、经济、政治、文化各方面的，虽在十个月抗战中有了某种程度的进步，但距离足以阻止敌之进攻及准备我之反攻的必要的程度，还远得很。且在量的方面，又不得不有所减弱。其各种有利因素，虽然都在起积极作用，但达到足以停止敌之进攻及准备我之反攻的程度则尚有待于巨大的努力。在国内，克服腐败现象，增加进步速度；在国外，克服助日势力，增加反日势力，尚非目前的现实。这一切，又规定了战争不能速胜，而只能是持久战。

持久战的三个阶段

（三五）中日战争既然是持久战，最后胜利又将是属已中国的，那末，就可以合理地设想，这种持久战，将具体地表现于三个阶段之中。第一个阶段，是敌之战略进攻、我之战略防御的时期。第二个阶段，是敌之战略保守、我之准备反攻的时期。第三个阶段，是我之战略反攻、敌之战略退却的时期。三个阶段的具体情况不能预断，但依目前条件来看，战争趋势中的某些大端是可以指出的。客观现实的行程将是异常丰富和曲折变化的，谁也不能造出一本中日战争的"流年"来；然而给战争趋势描画一个轮廓，却为战略指导所必需。所以，尽管描画的东西不能尽合将来的事实，而将为事实所校正，但是为着坚定地有目的地进行持久战的战略指导起见，描画轮廓的事仍然是需要的。

（三六）第一阶段，现在还未完结。敌之企图是攻占广州、武汉、兰州三点，并把三点联系起来。敌欲达此目的，至少出五十个师团，约一百五十万兵员，时间一年半至两年，用费将在一百万万日元以上。敌人如此深入，其困难是非常之大的，其后果将不堪设想。至欲完全占领粤汉铁路和西兰公路，将经历非常危险的战争，未必尽能达其企图。但是我们的作战计划，应把敌人可能占领三点甚至三点以外之某些部分地区并可能互相联系起来作为一种基础，部署持久战，即令敌如此做，我也有应付之方。这一阶段我所采取的战争形式，主要的是运动战，而以游击战和阵地战辅助之。阵地战虽在此阶段之第一期，由于国民党军事当局的主观错误把它放在主要地位，但从全阶段看，仍然是辅助的。此阶段中，中国已经结成了广大的统一战线，实现了空前的团结。敌虽已经采用过并且还将采用卑鄙无耻的劝降手段，企图不费大力实现其速决计划，整个地征服中国，但是过去的已经失败，今后的也难成功。此阶段中，中国虽有颇大的损失，但是同时却有颇大的进步，这种进步就成为第二阶段继续抗战的主要基础。此阶段中，苏联对于我国已经有了大量的援助。敌人方面，士气已开始表现颓靡，敌人陆军进攻的锐气，此阶段的中期已不如初期，末期将更不如初期。敌之财政和经济已开始表现其竭蹶状态，人民和士兵的厌战情绪已开始发生，战争指导集团的内部已开始表现其"战争的烦闷"，生长着对于战争前途的悲观。

（三七）第二阶段，可以名之曰战略的相持阶段。第一阶段之末尾，由于敌之兵力不足和我之坚强抵抗，敌人将不得不决定在一定限度上的战略进攻终点，到达此终点以后，即停止其战略进攻，转入保守占领地的阶段。此阶段内，敌之企图是保守占领地，以组织伪政府的欺骗办法据之为己有，而从中国人民身上尽量搜括东西，但是在他的面前又遇着顽强的游击战争。游击战争在第一阶段中乘着敌后空虚将有一个普遍的发展，建立许多根据地，基本上威胁到敌人占领地的保守，因此第二阶段仍将有广大的战争。此阶段中我之作战形式主要的是游击战，而以运动战辅助之。此时中国尚能保有大量的正规军，不过一方面因敌在其占领的大城市和大道中取战略守势，一方面因中国技术条件一时未能完备，尚难迅即举行战略反攻。除正面防御部队外，我军将大量地转入敌后，比较地分散配置，依托一切敌人未占区域，配合民众武装，向敌人占领地作广泛的和猛烈的游击战争，并尽可能地调动敌人于运动战中消灭之，如同现在山西的榜样。此阶段的战争是残酷的，地方将遇到严重的破坏。但是游击战争能够胜利，做得好，可能使敌只能保守占领地三分之一左右的区域，三分之二左右仍然是我们的，这就是敌人的大失败，中国的大胜利。那时，整个敌人占领地将分为三种地区：第一种是敌人的根据地，第二种是游击战争的根据地，第三种是双方争夺的游击区。这个阶段的时间的长短，依敌我力量增减变化的程度如何及国际形势变动如何而定，大体上我们要准备付给较长的时间，要熬得过这段艰难的路程。这将是中国很痛苦的时期，经济困难和汉奸捣乱将是两个很大的问题。敌人将大肆其破坏中国统一战线的活动，一切敌之占领地的汉奸组织将合流组成所谓"统一政府"。我们内部，因大城市的丧失和战争的困难，动摇分子将大倡其妥协论，悲观情绪将严重地增长。此时我们的任务，在于动员全国民众，齐心一致，绝不动摇地坚持战争，把统一战线扩大和巩固起来，排除一切悲观主义和妥协论，提倡艰苦斗争，实行新的战时政策，熬过这一段艰难的路程。此阶段内，必须号召全国坚决地维持一个统一政府，反对分裂，有计划地增强作战技术，改造军队，动员全民，准备反攻。此阶段中，国际形势将变到更于日本不利，虽可能有张伯伦[17]一类的迁就所谓"既成事实"的"现实主义"的调头出现，但主要的国际势力将变到进一步地援助中国。日本威胁南洋和威胁西伯利亚，将较之过去更加严重，甚至爆发新的战争。敌人方面，陷在中国泥潭中的几十个师团抽不出去。广大的游击战争和人民抗日运动将疲惫这一大批日本军，一方面大量地消耗之，又一方面进一步地增长其思乡厌战直至反战的心理，从精神上瓦解这个军队。日本在中国的掠夺虽然不能说它绝对不能有所成就，但是日本资本缺乏，又困于游击战争，急遽的大量的成就是不可能的。这个第二阶段是整个战争的过渡阶段，也将是最困难的时期，然而它是转变的枢纽。中国将变为独立国，还是沦为殖民地，不决定于第一阶段大城市之是否丧失，而决定于第二阶段全民族努力的程度。如能坚持抗战，坚持统一战线和坚持持久战，中国将在此阶段中获得转弱为强的力量。中国抗战的三幕戏，这是第二幕。由于全体演员的努力，最精彩的结幕便能很好地演出来。

（三八）第三阶段，是收复失地的反攻阶段。收复失地，主要地依靠中国自己在前阶段中准备着的和在本阶段中继续地生长着的力量。然而单只自己的力量还是不够的，还须依靠国际力量和敌国内部变化的援助，否则是不能胜利的，因此加重了中国的国际宣传和外交工作的任务。这个阶段，战争已不是战略防御，而将变为战略反攻了，在现象上，并将表现为战略进攻；已不是战略内线，而将逐渐地变为战略外线。直至打到鸭绿江边，才算结束了这个战争。第三阶段是持久战的最后阶段，所谓坚持战争到底，就是要走完这个阶段的全程。这个阶段我所采取的主要的战争形式仍将是运动战，但是阵地战将提到重要地位。如果说，第一阶段的阵地防御，由于当时的条件，不能看作重要的，那末，第三阶段的阵地攻击，由于条件的改变和任务的需要，将变成颇为重要的。此阶段内的游击战，仍将辅助运动战和阵地战而起其战略配合的作用，和第二阶段之变为主要形式者不相同。

（三九）这样看来，战争的长期性和随之而来的残酷性，是明显的。敌人不能整个地吞并中国，但是能够相当长期地占领中国的许多地方。中国也不能迅速地驱逐日本，但是大部分的土地将依然是中国的。最后是敌败我胜，但是必须经过一段艰难的路程。

（四〇）中国人民在这样长期和残酷的战争中间，将受到很好的锻炼。参加战争的各政党也将受到锻炼和考验。统一战线必须坚持下去；只有坚持统一战线，才能坚持战争；只有坚持统一战线和坚持战争，才能有最后胜利。果然是这样，一切困难就能够克服。跨过战争的艰难路程之后，胜利的坦途就到来了，这是战争的自然逻辑。

（四一）三个阶段中，敌我力量的变化将循着下述的道路前进。第一阶段敌是优势，我是劣势。我之这种劣势，须估计抗战以前到这一阶段末尾，有两种不同的变化。第一种是向下的。中国原来的劣势，经过第一阶段的消耗将更为严重，这就是土地、人口、经济力量、军事力量和文化机关等的减缩。第一阶段的末尾，也许要减缩到相当大的程度，特别是经济方面。这一点，将被人利用作为亡国论和妥协论的根据。然而必须看到第二种变化，即向上的变化。这就是战争中的经验，军队的进步，政治的进步，人民的动员，文化的新方向的发展，游击战争的出现，国际援助的增长等等。在第一阶段，向下的东西是旧的量和质，主要地表现在量上。向上的东西是新的量和质，主要地表现在质上。这第二种变化，就给了我们以能够持久和最后胜利的根据。

（四二）第一阶段中，敌人方面也有两种变化。第一种是向下的，表现在：几十万人的伤亡，武器和弹药的消耗，士气的颓靡，国内人心的不满，贸易的缩减，一百万万日元以上的支出，国际舆论的责备等等方面。这个方面，又给予我们以能够持久和最后胜利的根据。然而也要估计到敌人的第二种变化，即向上的变化。那就是他扩大了领土、人口和资源。在这点上面，又产生我们的抗战是持久战而不能速胜的根据，同时也将被一些人利用作为亡国论和妥协论的根据。但是我们必须估计敌人这种向上变化的暂时性和局部性。敌人是行将崩溃的帝国主义者，他占领中国的土地是暂时的。中国游击战争的猛烈发展，将使他的占领区实际上限制在狭小的地带。而且，敌人对中国土地的占领又产生了和加深了日本同外国的矛盾。再则，根据东三省的经验，日本在相当长的时间内，一般地只能是支出资本时期，不能是收获时期。所有这些，又是我们击破亡国论和妥协论而建立持久论和最后胜利论的根据。

（四三）第二阶段，上述双方的变化将继续发展，具体的情形不能预断，但是大体上将是日本继续向下，中国继续向上[18]。例如日本的军力、财力大量地消耗于中国的游击战争，国内人心更加不满，士气更加颓靡，国际更感孤立。中国则政治、军事、文化和人民动员将更加进步，游击战争更加发展，经济方面也将依凭内地的小工业和广大的农业而有某种程度的新发展，国际援助将逐渐地增进，将比现在的情况大为改观。这个第二阶段，也许将经过相当长的时间。在这个时间内，敌我力量对比将发生巨大的相反的变化，中国将逐渐上升，日本则逐渐下降。那时中国将脱出劣势，日本则脱出优势，先走到平衡的地位，再走到优劣相反的地位。然后中国大体上将完成战略反攻的准备而走到实行反攻、驱敌出国的阶段。应该重复地指出：所谓变劣势为优势和完成反攻准备，是包括中国自己力量的增长、日本困难的增长和国际援助的增长在内的，总合这些力量就能形成中国的优势，完成反攻的准备。

（四四）根据中国政治和经济不平衡的状态，第三阶段的战略反攻，在其前一时期将不是全国整齐划一的姿态，而是带地域性的和此起彼落的姿态。敌人采用各种分化手段破裂中国统一战线的企图，此阶段中并不会减弱，因此，中国内部团结的任务更加重要，务不令内部不调致战略反攻半途而废。此时期中，国际形势将变到大有利于中国。中国的任务，就在于利用这种国际形势取得自己的彻底解放，建立独立的民主国家，同时也就是帮助世界的反法西斯运动。

（四五）中国由劣势到平衡到优势，日本由优势到平衡到劣势，中国由防御到相持到反攻，日本由进攻到保守到退却——这就是中日战争的过程，中日战争的必然趋势。

（四六）于是问题和结论是：中国会亡吗？答复：不会亡，最后胜利是中国的。中国能够速胜吗？答复：不能速胜，必须是持久战。这个结论是正确的吗？我以为是正确的。

（四七）讲到这里，亡国论和妥协论者又将跑出来说：中国由劣势到平衡，需要有同日本相等的军力和经济力；由平衡到优势，需要有超过日本的军力和经济力；然而这是不可能的，因此上述结论是不正确的。

（四八）这就是所谓"唯武器论"，是战争问题中的机械论，是主观地和片面地看问题的意见。我们的意见与此相反，不但看到武器，而且看到人力。武器是战争的重要的因素，但不是决定的因素，决定的因素是人不是物。力量对比不但是军力和经济力的对比，而且是人力和人心的对比。军力和经济力是要人去掌握的。如果中国人的大多数、日本人的大多数、世界各国人的大多数是站在抗日战争方面的话，那末，日本少数人强制地掌握着的军力和经济力，还能算是优势吗？它不是优势，那末，掌握比较劣势的军力和经济力的中国，不就成了优势吗？没有疑义，中国只要坚持抗战和坚持统一战线，其军力和经济力是能够逐渐地加强的。而我们的敌人，经过长期战争和内外矛盾的削弱，其军力和经济力又必然要起相反的变化。在这种情况下，难道中国也不能变成优势吗？还不止此，目前我们不能把别国的军力和经济力大量地公开地算作自己方面的力量，难道将来也不能吗？如果日本的敌人不止中国一个，如果将来有一国或几国以其相当大量的军力和经济力公开地防御或攻击日本，公开地援助我们，那末，优势不更在我们一方面吗？日本是小国，其战争是退步的和野蛮的，其国际地位将益处于孤立；中国是大国，其战争是进步的和正义的，其国际地位将益处于多助。所有这些，经过长期发展，难道还不能使敌我优劣的形势确定地发生变化吗？

（四九）速胜论者则不知道战争是力量的竞赛，在战争双方的力量对比没有起一定的变化以前，就要举行战略的决战，就想提前到达解放之路，也是没有根据的。其意见实行起来，一定不免于碰壁。或者只是空谈快意，并不准备真正去做。最后则是事实先生跑将出来，给这些空谈家一瓢冷水，证明他们不过是一些贪便宜、想少费气力多得收成的空谈主义者。这种空谈主义过去和现在已经存在，但是还不算很多，战争发展到相持阶段和反攻阶段时，空谈主义可能多起来。但是在同时，如果第一阶段中国损失较大，第二阶段时间拖得很长，亡国论和妥协论更将大大地流行。所以我们的火力，应该主要地向着亡国论和妥协论方面，而以次要的火力，反对空谈主义的速胜论。

（五○）战争的长期性是确定了的，但是战争究将经过多少年月则谁也不能预断，这个完全要看敌我力量变化的程度才能决定。一切想要缩短战争时间的人们，惟有努力于增加自己力量减少敌人力量之一法。具体地说，惟有努力于作战多打胜仗，消耗敌人的军队，努力于发展游击战争，使敌之占领地限制于最小的范围，努力于巩固和扩大统一战线，团结全国力量，努力于建设新军和发展新的军事工业，努力于推动政治、经济和文化的进步，努力于工、农、商、学各界人民的动员，努力于瓦解敌军和争取敌军的士兵，努力于国际宣传争取国际的援助，努力于争取日本的人民及其它被压迫民族的援助，做了这一切，才能缩短战争的时间，此外不能有任何取巧图便的法门。

犬牙交错的战争

（五一）我们可以断言，持久战的抗日战争，将在人类战争史中表现为光荣的特殊的一页。犬牙交错的战争形态，就是颇为特殊的一点，这是由于日本的野蛮和兵力不足，中国的进步和土地广大这些矛盾因素产生出来的。犬牙交错的战争，在历史上也是有过的，俄国十月革命后的三年内战，就有过这种情形。但其在中国的特点，是其特殊的长期性和广大性，这将是突破历史纪录的东西。这种犬牙交错的形态，表现在下述的几种情况上。

（五二）内线和外线——抗日战争是整个处于内线作战的地位的；但是主力军和游击队的关系，则是主力军在内线，游击队在外线，形成夹攻敌人的奇观。各游击区的关系亦然。各个游击区都以自己为内线，而以其它各区为外线，又形成了很多夹攻敌人的火线。在战争的第一阶段，战略上内线作战的正规军是后退的，但是战略上外线作战的游击队则将广泛地向着敌人后方大踏步前进，第二阶段将更加猛烈地前进，形成了后退和前进的奇异形态。

（五三）有后方和无后方——利用国家的总后方，而把作战线伸至敌人占领地之最后限界的，是主力军。脱离总后方，而把作战线伸至敌后的，是游击队。但在每一游击区中，仍自有其小规模的后方，并依以建立非固定的作战线。和这个区别的，是每一游击区派遣出去向该区敌后临时活动的游击队，他们不但没有后方，也没有作战线。"无后方的作战"，是新时代中领土广大、人民进步、有先进政党和先进军队的情况之下的革命战争的特点，没有可怕而有大利，不应怀疑而应提倡。

（五四）包围和反包围——从整个战争看来，由于敌之战略进攻和外线作战，我处战略防御和内线作战地位，无疑我是在敌之战略包围中。这是敌对于我之第一种包围。由于我以数量上优势的兵力，对于从战略上的外线分数路向我前进之敌，采取战役和战斗上的外线作战方针，就可以把各路分进之敌的一路或几路放在我之包围中。这是我对于敌之第一种反包围。再从敌后游击战争的根据地看来，每一孤立的根据地都处于敌之四面或三面包围中，前者例如五台山，后者例如晋西北。这是敌对于我之第二种包围。但若将各个游击根据地联系起来看，并将各个游击根据地和正规军的阵地也联系起来看，我又把许多敌人都包围起来，例如在山西，我已三面包围了同蒲路（路之东西两侧及南端），四面包围了太原城；河北、山东等省也有许多这样的包围。这又是我对于敌之第二种反包围。这样，敌我各有加于对方的两种包围，大体上好似下围棋一样，敌对于我我对于敌之战役和战斗的作战，好似吃子，敌的据点（例如太原）和我之游击根据地（例如五台山），好似做眼。如果把世界性的围棋也算在内，那就还有第三种敌我包围，这就是侵略阵线与和平阵线的关系。敌以前者来包围中、苏、法、捷等国，我以后者反包围德、日、意。但是我之包围好似如来佛的手掌，它将化成一座横亘宇宙的五行山，把这几个新式孙悟空——法西斯侵略主义者，最后压倒在山底下，永世也不得翻身[19]。如果我能在外交上建立太平洋反日阵线，把中国作为一个战略单位，又把苏联及其它可能的国家也各作为一个战略单位，又把日本人民运动也作为一个战略单位，形成一个使法西斯孙悟空无处逃跑的天罗地网，那就是敌人死亡之时了。实际上，日本帝国主义完全打倒之日，必是这个天罗地网大体布成之时。这丝毫也不是笑话，而是战争的必然的趋势。

（五五）大块和小块——一种可能，是敌占地区将占中国本部之大半，而中国本部完整的区域只占一小半。这是一种情形。但是敌占大半中，除东三省等地外，实际只能占领大城市、大道和某些平地，依重要性说是一等的，依面积和人口来说可能只是敌占区中之小半，而普遍地发展的游击区，反居其大半。这又是一种情形。如果超越本部的范围，而把蒙古、新疆、青海、西藏算了进来，则在面积上中国未失地区仍然是大半，而敌占地区包括东三省在内，也只是小半。这又是一种情形。完整区域当然是重要的，应集大力去经营，不但政治、军事、经济等方面，文化方面也要紧。敌人已将我们过去的文化中心变为文化落后区域，而我们则要将过去的文化落后区域变为文化中心。同时，敌后广大游击区的经营也是非常之要紧的，也应把它们的各方面发展起来，也应发展其文化工作。总起来看，中国将是大块的乡村变为进步和光明的地区，而小块的敌占区，尤其是大城市，将暂时地变为落后和黑暗的地区。

（五六）这样看来，长期而又广大的抗日战争，是军事、政治、经济、文化各方面犬牙交错的战争，这是战争史上的奇观，中华民族的壮举，惊天动地的伟业。这个战争，不但将影响到中日两国，大大推动两国的进步，而且将影响到世界，推动各国首先是印度等被压迫民族的进步。全中国人都应自觉地投入这个犬牙交错的战争中去，这就是中华民族自求解放的战争形态，是半殖民地大国在二十世纪三十和四十年代举行的解放战争的特殊的形态。

为永久和平而战

（五七）中国抗日战争的持久性同争取中国和世界的永久和平，是不能分离的。没有任何一个历史时期像今天一样，战争是接近于永久和平的。由于阶级的出现，几千年来人类的生活中充满了战争，每一个民族都不知打了几多仗，或在民族集团之内打，或在民族集团之间打。打到资本主义社会的帝国主义时期，仗就打得特别广大和特别残酷。二十年前的第一次帝国主义大战，在过去历史上是空前的，但还不是绝后的战争。只有目前开始了的战争，接近于最后战争，就是说，接近于人类的永久和平。目前世界上已有三分之一的人口进入了战争，你们看，一个意大利，又一个日本，一个阿比西尼亚，又一个西班牙，再一个中国。参加战争的这些国家共有差不多六万万人口，几乎占了全世界总人口的三分之一。目前的战争的特点是无间断和接近永久和平的性质。为什么无间断？意大利同阿比西尼亚打了之后，接着意大利同西班牙打，德国也搭了股份，接着日本又同中国打。还要接着谁呢？无疑地要接着希特勒同各大国打。"法西斯主义就是战争"[20]，一点也不错。目前的战争发展到世界大战之间，是不会间断的，人类的战争灾难不可避免。为什么又说这次战争接近于永久和平？这次战争是在第一次世界大战所已开始的世界资本主义总危机发展的基础上发生的，由于这种总危机，逼使各资本主义国家走入新的战争，首先逼使各法西斯国家从事于新战争的冒险。我们可以预见这次战争的结果，将不是资本主义的获救，而是它的走向崩溃。这次战争，将比二十年前的战争更大，更残酷，一切民族将无可避免地卷入进去，战争时间将拖得很长，人类将遭受很大的痛苦。但是由于苏联的存在和世界人民觉悟程度的提高，这次战争中无疑将出现伟大的革命战争，用以反对一切反革命战争，而使这次战争带着为永久和平而战的性质。即使尔后尚有一个战争时期，但是已离世界的永久和平不远了。人类一经消灭了资本主义，便到达永久和平的时代，那时候便再也不要战争了。那时将不要军队，也不要兵船，不要军用飞机，也不要毒气。从此以后，人类将亿万斯年看不见战争。已经开始了的革命的战争，是这个为永久和平而战的战争的一部分。占着五万万以上人口的中日两国之间的战争，在这个战争中将占着重要的地位，中华民族的解放将从这个战争中得来。将来的被解放了的新中国，是和将来的被解放了的新世界不能分离的。因此，我们的抗日战争包含着为争取永久和平而战的性质。

（五八）历史上的战争分为两类，一类是正义的，一类是非正义的。一切进步的战争都是正义的，一切阻碍进步的战争都是非正义的。我们共产党人反对一切阻碍进步的非正义的战争，但是不反对进步的正义的战争。对于后一类战争，我们共产党人不但不反对，而且积极地参加。前一类战争，例如第一次世界大战，双方都是为着帝国主义利益而战，所以全世界的共产党人坚决地反对那一次战争。反对的方法，在战争未爆发前，极力阻止其爆发；既爆发后，只要有可能，就用战争反对战争，用正义战争反对非正义战争。日本的战争是阻碍进步的非正义的战争，全世界人民包括日本人民在内，都应该反对，也正在反对。我们中国，则从人民到政府，从共产党到国民党，一律举起了义旗，进行了反侵略的民族革命战争。我们的战争是神圣的、正义的，是进步的、求和平的。不但求一国的和平，而且求世界的和平，不但求一时的和平，而且求永久的和平。欲达此目的，便须决一死战，便须准备着一切牺牲，坚持到底，不达目的，决不停止。牺牲虽大，时间虽长，但是永久和平和永久光明的新世界，已经鲜明地摆在我们的前面。我们从事战争的信念，便建立在这个争取永久和平和永久光明的新中国和新世界的上面。法西斯主义和帝国主义要把战争延长到无尽期，我们则要把战争在一个不很久远的将来给以结束。为了这个目的，人类大多数应该拿出极大的努力。四亿五千万的中国人占了全人类的四分之一，如果能够一齐努力，打倒了日本帝国主义，创造了自由平等的新中国，对于争取全世界永久和平的贡献，无疑地是非常伟大的。这种希望不是空的，全世界社会经济的行程已经接近了这一点，只须加上多数人的努力，几十年工夫一定可以达到目的。

能动性在战争中

（五九）以上说的，都是说明为什么是持久战和为什么最后胜利是中国的，大体上都是说的"是什么"和"不是什么"。以下，将转到研究"怎样做"和"不怎样做"的问题上。怎样进行持久战和怎样争取最后胜利？这就是以下要答复的问题。为了这个，我们将依次说明下列的问题：能动性在战争中，战争和政治，抗战的政治动员，战争的目的，防御中的进攻，持久中的速决，内线中的外线，主动性，灵活性，计划性，运动战，游击战，阵地战，歼灭战，消耗战，乘敌之隙的可能性，抗日战争的决战问题，兵民是胜利之本。我们现在就从能动性问题说起吧。

（六○）我们反对主观地看问题，说的是一个人的思想，不根据和不符合于客观事实，是空想，是假道理，如果照了做去，就要失败，故须反对它。但是一切事情是要人做的，持久战和最后胜利没有人做就不会出现。做就必须先有人根据客观事实，引出思想、道理、意见，提出计划、方针、政策、战略、战术，方能做得好。思想等等是主观的东西，做或行动是主观见之于客观的东西，都是人类特殊的能动性。这种能动性，我们名之曰"自觉的能动性"，是人之所以区别于物的特点。一切根据和符合于客观事实的思想是正确的思想，一切根据于正确思想的做或行动是正确的行动。我们必须发扬这样的思想和行动，必须发扬这种自觉的能动性。抗日战争是要赶走帝国主义，变旧中国为新中国，必须动员全中国人民，统统发扬其抗日的自觉的能动性，才能达到目的。坐着不动，只有被灭亡，没有持久战，也没有最后胜利。

（六一）自觉的能动性是人类的特点。人类在战争中强烈地表现出这样的特点。战争的胜负，固然决定于双方军事、政治、经济、地理、战争性质、国际援助诸条件，然而不仅仅决定于这些；仅有这些，还只是有了胜负的可能性，它本身没有分胜负。要分胜负，还须加上主观的努力，这就是指导战争和实行战争，这就是战争中的自觉的能动性。

（六二）指导战争的人们不能超越客观条件许可的限度期求战争的胜利，然而可以而且必须在客观条件的限度之内，能动地争取战争的胜利。战争指挥员活动的舞台，必须建筑在客观条件的许可之上，然而他们凭借这个舞台，却可以导演出很多有声有色、威武雄壮的戏剧来。在既定的客观物质的基础之上，抗日战争的指挥员就要发挥他们的威力，提挈全军，去打倒那些民族的敌人，改变我们这个被侵略被压迫的社会国家的状态，造成自由平等的新中国，这里就用得着而且必须用我们的主观指导的能力。我们不赞成任何一个抗日战争的指挥员，离开客观条件，变为乱撞乱碰的鲁莽家，但是我们必须提倡每个抗日战争的指挥员变为勇敢而明智的将军。他们不但要有压倒敌人的勇气，而且要有驾驭整个战争变化发展的能力。指挥员在战争的大海中游泳，他们要不使自己沉没，而要使自己决定地有步骤地到达彼岸。作为战争指导规律的战略战术，就是战争大海中的游泳术。

战争和政治

（六三）"战争是政治的继续"，在这点上说，战争就是政治，战争本身就是政治性质的行动，从古以来没有不带政治性的战争。抗日战争是全民族的革命战争，它的胜利，离不开战争的政治目的——驱逐日本帝国主义、建立自由平等的新中国，离不开坚持抗战和坚持统一战线的总方针，离不开全国人民的动员，离不开官兵一致、军民一致和瓦解敌军等项政治原则，离不开统一战线政策的良好执行，离不开文化的动员，离不开争取国际力量和敌国人民援助的努力。一句话，战争一刻也离不了政治。抗日军人中，如有轻视政治的倾向，把战争孤立起来，变为战争绝对主义者，那是错误的，应加纠正。

（六四）但是战争有其特殊性，在这点上说，战争不即等于一般的政治。"战争是政治的特殊手段的继续"[21]。政治发展到一定的阶段，再也不能照旧前进，于是爆发了战争，用以扫除政治道路上的障碍。例如中国的半独立地位，是日本帝国主义政治发展的障碍，日本要扫除它，所以发动了侵略战争。中国呢？帝国主义压迫，早就是中国资产阶级民主革命的障碍，所以有了很多次的解放战争，企图扫除这个障碍。日本现在用战争来压迫，要完全断绝中国革命的进路，所以不得不举行抗日战争，决心要扫除这个障碍。障碍既除，政治的目的达到，战争结束。障碍没有扫除得干净，战争仍须继续进行，以求贯彻。例如抗日的任务未完，有想求妥协的，必不成功；因为即使因某种缘故妥协了，但是战争仍要起来，广大人民必定不服，必要继续战争，贯彻战争的政治目的。因此可以说，政治是不流血的战争，战争是流血的政治。

（六五）基于战争的特殊性，就有战争的一套特殊组织，一套特殊方法，一种特殊过程。这组织，就是军队及其附随的一切东西。这方法，就是指导战争的战略战术。这过程，就是敌对的军队互相使用有利于己不利于敌的战略战术从事攻击或防御的一种特殊的社会活动形态。因此，战争的经验是特殊的。一切参加战争的人们，必须脱出寻常习惯，而习惯于战争，方能争取战争的胜利。

抗日的政治动员

（六六）如此伟大的民族革命战争，没有普遍和深入的政治动员，是不能胜利的。抗日以前，没有抗日的政治动员，这是中国的大缺陷，已经输了敌人一着。抗日以后，政治动员也非常之不普遍，更不说深入。人民的大多数，是从敌人的炮火和飞机炸弹那里听到消息的。这也是一种动员，但这是敌人替我们做的，不是我们自己做的。偏远地区听不到炮声的人们，至今还是静悄悄地在那里过活。这种情形必须改变，不然，拚死活的战争就得不到胜利。决不可以再输敌人一着，相反，要大大地发挥这一着去制胜敌人。这一着是关系绝大的；武器等等不如人尚在其次，这一着实在是头等重要。动员了全国的老百姓，就造成了陷敌于灭顶之灾的汪洋大海，造成了弥补武器等等缺陷的补救条件，造成了克服一切战争困难的前提。要胜利，就要坚持抗战，坚持统一战线，坚持持久战。然而一切这些，离不开动员老百姓。要胜利又忽视政治动员，叫做"南其辕而北其辙"，结果必然取消了胜利。

（六七）什么是政治动员呢？首先是把战争的政治目的告诉军队和人民。必须使每个士兵每个人民都明白为什么要打仗，打仗和他们有什么关系。抗日战争的政治目的是"驱逐日本帝国主义，建立自由平等的新中国"，必须把这个目的告诉一切军民人等，方能造成抗日的热潮，使几万万人齐心一致，贡献一切给战争。其次，单单说明目的还不够，还要说明达到此目的的步骤和政策，就是说，要有一个政治纲领。现在已经有了《抗日救国十大纲领》[22]，又有了一个《抗战建国纲领》[23]，应把它们普及于军队和人民，并动员所有的军队和人民实行起来。没有一个明确的具体的政治纲领，是不能动员全军全民抗日到底的。其次，怎样去动员？靠口说，靠传单布告，靠报纸书册，靠戏剧电影，靠学校，靠民众团体，靠干部人员。现在国民党统治地区有的一些，沧海一粟，而且方法不合民众口味，神气和民众隔膜，必须切实地改一改。其次，不是一次动员就够了，抗日战争的政治动员是经常的。不是将政治纲领背诵给老百姓听，这样的背诵是没有人听的；要联系战争发展的情况，联系士兵和老百姓的生活，把战争的政治动员，变成经常的运动。这是一件绝大的事，战争首先要靠它取得胜利。

战争的目的

（六八）这里不是说战争的政治目的，抗日战争的政治目的是"驱逐日本帝国主义，建立自由平等的新中国"，前面已经说过了。这里说的，是作为人类流血的政治的所谓战争，两军相杀的战争，它的根本目的是什么。战争的目的不是别的，就是"保存自己，消灭敌人"（消灭敌人，就是解除敌人的武装，也就是所谓"剥夺敌人的抵抗力"，不是要完全消灭其肉体）。古代战争，用矛用盾：矛是进攻的，为了消灭敌人；盾是防御的，为了保存自己。直到今天的武器，还是这二者的继续。轰炸机、机关枪、远射程炮、毒气，是矛的发展；防空掩蔽部、钢盔、水泥工事、防毒面具，是盾的发展。坦克，是矛盾二者结合为一的新式武器。进攻，是消灭敌人的主要手段，但防御也是不能废的。进攻，是直接为了消灭敌人的，同时也是为了保存自己，因为如不消灭敌人，则自己将被消灭。防御，是直接为了保存自己的，但同时也是辅助进攻或准备转入进攻的一种手段。退却，属于防御一类，是防御的继续；而追击，则是进攻的继续。应该指出：战争目的中，消灭敌人是主要的，保存自己是第二位的，因为只有大量地消灭敌人，才能有效地保存自己。因此，作为消灭敌人之主要手段的进攻是主要的，而作为消灭敌人之辅助手段和作为保存自己之一种手段的防御，是第二位的。战争实际中，虽有许多时候以防御为主，而在其余时候以进攻为主，然而通战争的全体来看，进攻仍然是主要的。

（六九）怎样解释战争中提倡勇敢牺牲呢？岂非与"保存自己"相矛盾？不相矛盾，是相反相成的。战争是流血的政治，是要付代价的，有时是极大的代价。部分的暂时的牺牲（不保存），为了全体的永久的保存。我们说，基本上为着消灭敌人的进攻手段中，同时也含了保存自己的作用，理由就在这里。防御必须同时有进攻，而不应是单纯的防御，也是这个道理。

（七○）保存自己消灭敌人这个战争的目的，就是战争的本质，就是一切战争行动的根据，从技术行动起，到战略行动止，都是贯彻这个本质的。战争目的，是战争的基本原则，一切技术的、战术的、战役的、战略的原理原则，一点也离不开它。射击原则的"荫蔽身体，发扬火力"是什么意思呢？前者为了保存自己，后者为了消灭敌人。因为前者，于是利用地形地物，采取跃进运动，疏开队形，种种方法都发生了。因为后者，于是扫清射界，组织火网，种种方法也发生了。战术上的突击队、钳制队、预备队，第一种为了消灭敌人，第二种为了保存自己，第三种准备依情况使用于两个目的——或者增援突击队，或者作为追击队，都是为了消灭敌人；或者增援钳制队，或者作为掩护队，都是为了保存自己。照这样，一切技术、战术、战役、战略原则，一切技术、战术、战役、战略行动，一点也离不开战争的目的，它普及于战争的全体，贯彻于战争的始终。

（七一）抗日战争的各级指导者，不能离开中日两国之间各种互相对立的基本因素去指导战争，也不能离开这个战争目的去指导战争。两国之间各种互相对立的基本因素展开于战争的行动中，就变成互相为了保存自己消灭敌人而斗争。我们的战争，在于力求每战争取不论大小的胜利，在于力求每战解除敌人一部分武装，损伤敌人一部分人马器物。把这些部分地消灭敌人的成绩积累起来，成为大的战略胜利，达到最后驱敌出国，保卫祖国，建设新中国的政治目的。

防御中的进攻，持久中的速决，内线中的外线

（七二）现在来研究抗日战争中的具体的战略方针。我们已说过了，抗日的战略方针是持久战，是的，这是完全对的。但这是一

般的方针，还不是具体的方针。怎样具体地进行持久战呢？这就是我们现在要讨论的问题。我们的答复是：在第一和第二阶段即敌之进攻和保守阶段中，应该是战略防御中的战役和战斗的进攻战，战略持久中的战役和战斗的速决战，战略内线中的战役和战斗的外线作战。在第三阶段中，应该是战略的反攻战。

（七三）由于日本是帝国主义的强国，我们是半殖民地半封建的弱国，日本是采取战略进攻方针的，我们则居于战略防御地位。日本企图采取战略的速决战，我们应自觉地采取战略的持久战。日本用其战斗力颇强的几十个师团的陆军（目前已到了三十个师团）和一部分海军，从陆海两面包围和封锁中国，又用空军轰炸中国。目前日本的陆军已占领从包头到杭州的长阵线，海军则到了福建广东，形成了大范围的外线作战。我们则处于内线作战地位。所有这些，都是由敌强我弱这个特点造成的。这是一方面的情形。

（七四）然而在另一方面，则适得其反。日本虽强，但兵力不足。中国虽弱，但地大、人多、兵多。这里就产生了两个重要的结果。第一，敌以少兵临大国，就只能占领一部分大城市、大道和某些平地。由是，在其占领区域，则空出了广大地面无法占领，这就给了中国游击战争以广大活动的地盘。在全国，即使敌能占领广州、武汉、兰州之线及其附近的地区，但以外的地区是难于占领的，这就给了中国以进行持久战和争取最后胜利的总后方和中枢根据地。第二，敌以少兵临多兵，便处于多兵的包围中。敌分路向我进攻，敌处战略外线，我处战略内线，敌是战略进攻，我是战略防御，看起来我是很不利的。然而我可以利用地广和兵多两个长处，不作死守的阵地战，采用灵活的运动战，以几个师对他一个师，几万人对他一万人，几路对他一路，从战场的外线，突然包围其一路而攻击之。于是敌之战略作战上的外线和进攻，在战役和战斗的作战上，就不得不变成内线和防御。我之战略作战上的内线和防御，在战役和战斗的作战上就变成了外线和进攻。对其一路如此，对其它路也是如此。以上两点，都是从敌小我大这一特点发生的。又由于敌兵虽少，乃是强兵（武器和人员的教养程度），我兵虽多，乃是弱兵（也仅是武器和人员的教养程度，不是士气），因此，在战役和战斗的作战上，我不但应以多兵打少兵，从外线打内线，还须采取速决战的方针。为了实行速决，一般应不打驻止中之敌，而打运动中之敌。我预将大兵荫蔽集结于敌必经通路之侧，乘敌运动之际，突然前进，包围而攻击之，打他一个措手不及，迅速解决战斗。打得好，可能全部或大部或一部消灭他；打不好，也给他一个大的杀伤。一战如此，他战皆然。不说多了，每个月打得一个较大的胜仗，如像平型关台儿庄一类的，就能大大地沮丧敌人的精神，振起我军的士气，号召世界的声援。这样，我之战略的持久战，到战场作战就变成速决战了。敌之战略的速决战，经过许多战役和战斗的败仗，就不得不改为持久战。

（七五）上述这样的战役和战斗的作战方针，一句话说完，就是："外线的速决的进攻战"。这对于我之战略方针"内线的持久的防御战"说来，是相反的；然而，又恰是实现这样的战略方针之必要的方针。如果战役和战斗方针也同样是"内线的持久的防御战"，例如抗战初起时期之所为，那就完全不适合敌小我大、敌强我弱这两种情况，那就决然达不到战略目的，达不到总的持久战，而将为敌人所击败。所以，我们历来主张全国组成若干个大的野战兵团，其兵力针对着敌人每个野战兵团之兵力而二倍之、三倍之或四倍之，采用上述方针，与敌周旋于广大战场之上。这种方针，不但是正规战争用得着，游击战争也用得着，而且必须要用它。不但适用于战争的某一阶段，而且适用于战争的全过程。战略反攻阶段，我之技术条件增强，以弱敌强这种情况即使完全没有了，我仍用多兵从外线采取速决的进攻战，就更能收大批俘获的成效。例如我用两个或三个或四个机械化的师对敌一个机械化的师，更能确定地消灭这个师。几个大汉打一个大汉之容易打胜，这是常识中包含的真理。

（七六）如果我们坚决地采取了战场作战的"外线的速决的进攻战"，就不但在战场上改变着敌我之间的强弱优劣形势，而且将逐渐地变化着总的形势。在战场上，因为我是进攻，敌是防御；我是多兵处外线，敌是少兵处内线；我是速决，敌虽企图持久待援，但不能由他作主；于是在敌人方面，强者就变成了弱者，优势就变成了劣势；我军方面反之，弱者变成了强者，劣势变成了优势。在打了许多这样的胜仗之后，总的敌我形势便将引起变化。这就是说，集合了许多战场作战的外线的速决的进攻战的胜利以后，就逐渐地增强了自己，削弱了敌人，于是总的强弱优劣形势，就不能不受其影响而发生变化。到那时，配合着我们自己的其它条件，再配合着敌人内部的变动和国际上的有利形势，就能使敌我总的形势走到平衡，再由平衡走到我优敌劣。那时，就是我们实行反攻驱敌出国的时机了。

（七七）战争是力量的竞赛，但力量在战争过程中变化其原来的形态。在这里，主观的努力，多打胜仗，少犯错误，是决定的因素。客观因素具备着这种变化的可能性，但实现这种可能性，就需要正确的方针和主观的努力。这时候，主观作用是决定的了。

主动性，灵活性，计划性

（七八）上面说过的战役和战斗的外线的速决的进攻战，中心点在于一个进攻；外线是说的进攻的范围，速决是说的进攻的时间，所以叫它做"外线的速决的进攻战"。这是实行持久战的最好的方针，也即是所谓运动战的方针。但是这个方针实行起来，离不了主动性、灵活性和计划性。我们现在就来研究这三个问题。

（七九）前面已说过了自觉的能动性，为什么又说主动性呢？自觉的能动性，说的是自觉的活动和努力，是人之所以区别于物的特点，这种人的特点，特别强烈地表现于战争中，这些是前面说过了的。这里说的主动性，说的是军队行动的自由权，是用以区别于被迫处于不自由状态的。行动自由是军队的命脉，失了这种自由，军队就接近于被打败或被消灭。一个士兵被缴械，是这个士兵失了行动自由被迫处于被动地位的结果。一个军队的战败，也是一样。为此缘故，战争的双方，都力争主动，力避被动。我们提出的外线的速决的进攻战，以及为了实现这种进攻战的灵活性、计划性，可以说都是为了争取主动权，以便逼敌处于被动地位，达到保存自己消灭敌人之目的。但主动或被动是和战争力量的优势或劣势分不开的。因而也是和主观指导的正确或错误分不开的。此外，也还有利用敌人的错觉和不意来争取自己主动和逼敌处于被动的情形。下面就来分析这几点。

（八〇）主动是和战争力量的优势不能分离的，而被动则和战争力量的劣势分不开。战争力量的优势或劣势，是主动或被动的客观基础。战略的主动地位，自然以战略的进攻战为较能掌握和发挥，然而贯彻始终和普及各地的主动地位，即绝对的主动权，只有以绝对优势对绝对劣势才有可能。一个身体壮健者和一个重病患者角斗，前者便有绝对的主动权。如果日本没有许多不可克服的矛盾，例如它能一下出几百万至一千万大兵，财源比现在多过几倍，又没有民众和外国的敌对，又不实行野蛮政策招致中国人民拚死命反抗，那它便能保持一种绝对的优势，它便有一种贯彻始终和普及各地的绝对的主动权。但在历史上，这类绝对优势的事情，在战争和战役的结局是存在的，战争和战役的开头则少见。例如在第一次世界大战中，德国屈服的前夜，这时协约国变成了绝对优势，德国则变成了绝对劣势，结果德国失败，协约国获胜，这是战争结局存在着绝对的优势和劣势之例。又如台儿庄胜利的前夜，这时当地孤立的日军经过苦战之后，已处于绝对的劣势，我军则造成了绝对的优势，结果敌败我胜，这是战役结局存在着绝对的优势和劣势之例。战争或战役也有以相对的优劣或平衡状态而结局的，那时，在战争则出现妥协，在战役则出现对峙。但一般是以绝对的优劣而分胜负居多数。所有这些，都是战争或战役的结局，而非战争或战役的开头。中日战争的最后结局，可以预断，日本将以绝对劣势而失败，中国将以绝对优势而获胜；但是在目前，则双方的优劣都不是绝对的而是相对的。日本因其具有强的军力、经济力和政治组织力这个有利因素，对于我们弱的军力、经济力和政治组织力，占了优势，因而造成了它的主动权的基础。但是因为它的军力等等数量不多，又有其它许多不利因素，它的优势便为它自己的矛盾所减杀。及到中国，又碰到了中国的地大、人多、兵多和坚强的民族抗战，它的优势再为之减杀。于是在总的方面，它的地位就变成一种相对的优势，因而其主动权的发挥和维持就受了限制，也成了相对的东西。中国方面，虽然在力量的强度上是劣势，因此造成了战略上的某种被动姿态，但是在地理、人口和兵员的数量上，并且又在人民和军队的敌忾心和士气上，却处于优势，这种优势再加上其它的有利因素，便减杀了自己军力、经济力等的劣势的程度，使之变为战略上的相对的劣势。因而也减少了被动的程度，仅处于战略上的相对的被动地位。然而被动总是不利的，必须力求脱离它。军事上的办法，就是坚决地实行外线的速决的进攻战和发动敌后的游击战争，在战役的运动战和游击战中取得许多局部的压倒敌人的优势和主动地位。通过这样许多战役的局部优势和局部主动地位，就能逐渐地造成战略的优势和战略的主动地位，战略的劣势和被动地位就能脱出了。这就是主动和被动之间、优势和劣势之间的相互关系。

（八一）由此也就可以明白主动或被动和主观指导之间的关系。如上所述，我之相对的战略劣势和战略被动地位，是能够脱出的，方法就是人工地造成我们许多的局部优势和局部主动地位，去剥夺敌人的许多局部优势和局部主动地位，把他抛入劣势和被动。把这些局部的东西集合起来，就成了我们的战略优势和战略主动，敌人的战略劣势和战略被动。这样的转变，依靠主观上的正确指导。

为什么呢？我要优势和主动，敌人也要这个，从这点上看，战争就是两军指挥员以军力财力等项物质基础作地盘，互争优势和主动的主观能力的竞赛。竞赛结果，有胜有败，除了客观物质条件的比较外，胜者必由于主观指挥的正确，败者必由于主观指挥的错误。我们承认战争现象是较之任何别的社会现象更难捉摸，更少确实性，即更带所谓"盖然性"。但战争不是神物，仍是世间的一种必然运动，因此，孙子的规律，"知彼知己，百战不殆"[24]，仍是科学的真理。错误由于对彼己的无知，战争的特性也使人们在许多的场合无法全知彼己，因此产生了战争情况和战争行动的不确实性，产生了错误和失败。然而不管怎样的战争情况和战争行动，知其大略，知其要点，是可能的。先之以各种侦察手段，继之以指挥员的聪明的推论和判断，减少错误，实现一般的正确指导，是做得到的。我们有了这个"一般地正确的指导"做武器，就能多打胜仗，就能变劣势为优势，变被动为主动。这是主动或被动和主观指导的正确与否之间的关系。

（八二）主观指导的正确与否，影响到优势劣势和主动被动的变化，观于强大之军打败仗、弱小之军打胜仗的历史事实而益信。中外历史上这类事情是多得很的。中国如晋楚城濮之战[25]，楚汉成皋之战[26]，韩信破赵之战[27]，新汉昆阳之战[28]，袁曹官渡之战[29]，吴魏赤壁之战[30]，吴蜀彝陵之战[31]，秦晋淝水之战[32]等等，外国如拿破仑的多数战役[33]，十月革命后的苏联内战，都是以少击众，以劣势对优势而获胜。都是先以自己局部的优势和主动，向着敌人局部的劣势和被动，一战而胜，再及其余，各个击破，全局因而转成了优势，转成了主动。在原占优势和主动之敌则反是；由于其主观错误和内部矛盾，可以将其很好的或较好的优势和主动地位，完全丧失，化为败军之将，亡国之君。由此可知，战争力量的优劣本身，固然是决定主动或被动的客观基础，但还不是主动或被动的现实事物，必待经过斗争，经过主观能力的竞赛，方才出现事实上的主动或被动。在斗争中，由于主观指导的正确或错误，可以化劣势为优势，化被动为主动；也可以化优势为劣势，化主动为被动。一切统治王朝打不赢革命军，可见单是某种优势还没有确定主动地位，更没有确定最后胜利。主动和胜利，是可以根据真实的情况，经过主观能力的活跃，取得一定的条件，而由劣势和被动者从优势和主动者手里夺取过来的。

　　（八三）错觉和不意，可以丧失优势和主动。因而有计划地造成敌人的错觉，给以不意的攻击，是造成优势和夺取主动的方法，而且是重要的方法。错觉是什么呢？"八公山上，草木皆兵"[34]，是错觉之一例。"声东击西"，是造成敌人错觉之一法。在优越的民众条件具备，足以封锁消息时，采用各种欺骗敌人的方法，常能有效地陷敌于判断错误和行动错误的苦境，因而丧失其优势和主动。"兵不厌诈"，就是指的这件事情。什么是不意？就是无准备。优势而无准备，不是真正的优势，也没有主动。懂得这一点，劣势而有准备之军，常可对敌举行不意的攻势，把优势者打败。我们说运动之敌好打，就是因为敌在不意即无准备中。这两件事——造成敌人的错觉和出以不意的攻击，即是以战争的不确实性给予敌人，而给自己以尽可能大的确实性，用以争取我之优势和主动，争取我之胜利。要做到这些，先决条件是优越的民众组织。因此，发动所有一切反对敌人的老百姓，一律武装起来，对敌进行广泛的袭击，同时即用以封锁消息，掩护我军，使敌无从知道我军将在什么地方什么时候去攻击他，造成他的错觉和不意的客观基础，是非常之重要的。过去土地革命战争时代的中国红军，以弱小的军力而常打胜仗，得力于组织起来和武装起来了的民众是非常之大的。民族战争照规矩应比土地革命战争更能获得广大民众的援助；可是因为历史的错误[35]，民众是散的，不但仓卒难为我用，且时为敌人所利用。只有坚决地广泛地发动全体的民众，方能在战争的一切需要上给以无穷无尽的供给。在这个给敌以错觉和给敌以不意以便战而胜之的战争方法上，也就一定能起大的作用。我们不是宋襄公，不要那种蠢猪式的仁义道德[36]。我们要把敌人的眼睛和耳朵尽可能地封住，使他们变成瞎子和聋子，要把他们的指挥员的心尽可能地弄得混乱些，使他们变成疯子，用以争取自己的胜利。所有这些，也都是主动或被动和主观指导之间的相互关系。战胜日本是少不了这种主观指导的。

（八四）大抵日本在其进攻阶段中，因其军力之强和利用我之主观上的历史错误和现时错误，它是一般地处于主动地位的。但是这种主动，已随其本身带着许多不利因素及其在战争中也犯了些主观错误（详论见后），与乎我方具备着许多有利因素，而开始了部分的减弱。敌之在台儿庄失败和山西困处，就是显证。我在敌后游击战争的广大发展，则使其占领地的守军完全处于被动地位。虽则敌人此时还在其主动的战略进攻中，但他的主动将随其战略进攻的停止而结束。敌之兵力不足，没有可能作无限制的进攻，这是他不能继续保持主动地位的第一个根源。我之战役的进攻战，在敌后的游击战争及其它条件，这是他不能不停止进攻于一定限度和不能继续保持主动地位的第二个根源。苏联的存在及其它国际变化，是第三个根源。由此可见，敌人的主动地位是有限制的，也是能够破坏的。中国如能在作战方法上坚持主力军的战役和战斗的进攻战，猛烈地发展敌后的游击战争，并从政治上大大地发动民众，我之战略主动地位便能逐渐树立起来。

（八五）现在来说灵活性。灵活性是什么呢？就是具体地实现主动性于作战中的东西，就是灵活地使用兵力。灵活地使用兵力这件事，是战争指挥的中心任务，也是最不容易做好的。战争的事业，除了组织和教育军队，组织和教育人民等项之外，就是使用军队于战斗，而一切都是为了战斗的胜利。组织军队等等固然困难，但使用军队则更加困难，特别是在以弱敌强的情况之中。做这件事需要极大的主观能力，需要克服战争特性中的纷乱、黑暗和不确实性，而从其中找出条理、光明和确实性来，方能实现指挥上的灵活性。

（八六）抗日战争战场作战的基本方针，是外线的速决的进攻战。执行这个方针，有兵力的分散和集中、分进和合击、攻击和防御、突击和钳制、包围和迂回、前进和后退种种的战术或方法。懂得这些战术是容易的，灵活地使用和变换这些战术，就不容易了。这里有时机、地点、部队三个关节。不得其时，不得其地，不得于部队之情况，都将不能取胜。例如进攻某一运动中之敌，打早了，暴露了自己，给了敌人以预防条件；打迟了，敌已集中驻止，变为啃硬骨头。这就是时机问题。突击点选在左翼，恰当敌之弱点，容易取胜；选在右翼，碰在敌人的钉子上，不能奏效。这就是地点问题。以我之某一部队执行某种任务，容易取胜；以另一部队执行同样任务，难于收效。这就是部队情况问题。不但使用战术，还须变换战术。攻击变为防御，防御变为攻击，前进变为后退，后退变为前进，钳制队变为突击队，突击队变为钳制队，以及包围迂回等等之互相变换，依据敌我部队、敌我地形的情况，及时地恰当地给以变换，是灵活性的指挥之重要任务。战斗指挥如此，战役和战略指挥也是如此。

（八七）古人所谓"运用之妙，存乎一心"[37]，这个"妙"，我们叫做灵活性，这是聪明的指挥员的出产品。灵活不是妄动，妄动是应该拒绝的。灵活，是聪明的指挥员，基于客观情况，"审时度势"（这个势，包括敌势、我势、地势等项）而采取及时的和恰当的处置方法的一种才能，即是所谓"运用之妙"。基于这种运用之妙，外线的速决的进攻战就能较多地取得胜利，就能转变敌我优劣形势，就能实现我对于敌的主动权，就能压倒敌人而击破之，而最后胜利就属于我们了。

（八八）现在来说计划性。由于战争所特有的不确实性，实现计划性于战争，较之实现计划性于别的事业，是要困难得多的。然而，"凡事预则立，不预则废"[38]，没有事先的计划和准备，就不能获得战争的胜利。战争没有绝对的确实性，但不是没有某种程度的相对的确实性。我之一方是比较地确实的。敌之一方很不确实，但也有朕兆〔注：现通常写为"征兆"。——中文马克思主义文库〕可寻，有端倪可察，有前后现象可供思索。这就构成了所谓某种程度的相对的确实性，战争的计划性就有了客观基础。近代技术（有线电、无线电、飞机、汽车、铁道、轮船等）的发达，又使战争的计划性增大了可能。但由于战争只有程度颇低和时间颇暂的确实性，战争的计划性就很难完全和固定，它随战争的运动（或流动，或推移）而运动，且依战争范围的大小而有程度的不同。战术计划，例如小兵团和小部队的攻击或防御计划，常须一日数变。战役计划，即大兵团的行动计划，大体能终战役之局，但在该战役内，部分的改变是常有的，全部的改变也间或有之。战略计划，是基于战争双方总的情况而来的，有更大的固定的程度，但也只在一定的战略阶段内适用，战争向着新的阶段推移，战略计划便须改变。战术、战役和战略计划之各依其范围和情况而确定而改变，是战争指挥的重要关节，也即是战争灵活性的具体的实施，也即是实际的运用之妙。抗日战争的各级指挥员，对此应当加以注意。

（八九）有些人，基于战争的流动性，就从根本上否认战争计划或战争方针之相对的固定性，说这样的计划或方针是"机械的"东西。这种意见是错误的。如上条所述，我们完全承认：由于战争情况之只有相对的确实性和战争是迅速地向前流动的（或运动的，推移的），战争的计划或方针，也只应给以相对的固定性，必须根据情况的变化和战争的流动而适时地加以更换或修改，不这样做，我们就变成机械主义者。然而决不能否认一定时间内的相对地固定的战争计划或方针；否认了这点，就否认了一切，连战争本身，连说话的人，都否认了。由于战争的情况和行动都有其相对的固定性，因而应之而生的战争计划或方针，也就必须拿相对的固定性赋予它。例如，由于华北战争的情况和八路军分散作战的行动有其在一定阶段内的固定性，因而在这一定阶段内赋予相对的固定性于八路军的"基本的是游击战，但不放松有利条件下的运动战"这种战略的作战方针，是完全必要的。战役方针，较之上述战略方针适用的时间要短促些，战术方针更加短促，然而都有其一定时间的固定性。否认了这点，战争就无从着手，成为毫无定见，这也不是、那也不是，或者这也是、那也是的战争相对主义了。没有人否认，就是在某一定时间内适用的方针，它也是在流动的，没有这种流动，就不会有这一方针的废止和另一方针的采用。然而这种流动是有限制的，即流动于执行这一方针的各种不同的战争行动的范围中，而不是这一方针的根本性质的流动，即是说，是数的流动，不是质的流动。这种根本性质，在一定时间内是决不流动的，我们所谓一定时间内的相对的固定性，就是指的这一点。在绝对流动的整个战争长河中有其各个特定阶段上的相对的固定性——这就是我们对于战争计划或战争方针的根本性质的意见。

（九○）在说过了战略上的内线的持久的防御战和战役战斗上的外线的速决的进攻战，又说过了主动性、灵活性和计划性之后，我们可以总起来说几句。抗日战争应该是有计划的。战争计划即战略战术的具体运用，要带灵活性，使之能适应战争的情况。要处处照顾化劣势为优势，化被动为主动，以便改变敌我之间的形势。而一切这些，都表现于战役和战斗上的外线的速决的进攻战，同时也就表现于战略上的内线的持久的防御战之中。

运动战，游击战，阵地战

（九一）作为战争内容的战略内线、战略持久、战略防御中的战役和战斗上的外线的速决的进攻战，在战争形式上就表现为运动战。运动战，就是正规兵团在长的战线和大的战区上面，从事于战役和战斗上的外线的速决的进攻战的形式。同时，也把为了便利于执行这种进攻战而在某些必要时机执行着的所谓"运动性的防御"包括在内，并且也把起辅助作用的阵地攻击和阵地防御包括在内。它的特点是：正规兵团，战役和战斗的优势兵力，进攻性和流动性。

（九二）中国版图广大，兵员众多，但军队的技术和教养不足；敌人则兵力不足，但技术和教养比较优良。在此种情形下，无疑地应以进攻的运动战为主要的作战形式，而以其它形式辅助之，组成整个的运动战。在这里，要反对所谓"有退无进"的逃跑主义，同时也要反对所谓"有进无退"的拚命主义。

（九三）运动战的特点之一，是其流动性，不但许可而且要求野战军的大踏步的前进和后退。然而，这和韩复榘式的逃跑主义[39]是没有相同之点的。战争的基本要求是：消灭敌人；其另一要求是：保存自己。保存自己的目的，在于消灭敌人；而消灭敌人，又是保存自己的最有效的手段。因此，运动战决不能被韩复榘一类人所借口，决不是只有向后的运动，没有向前的运动；这样的"运动"，否定了运动战的基本的进攻性，实行的结果，中国虽大，也是要被"运动"掉的。

（九四）然而另一种思想也是不对的，即所谓有进无退的拚命主义。我们主张以战役和战斗上的外线的速决的进攻战为内容的运动战，其中包括了辅助作用的阵地战，又包括了"运动性的防御"和退却，没有这些，运动战便不能充分地执行。拚命主义是军事上的近视眼，其根源常是惧怕丧失土地。拚命主义者不知道运动战的特点之一是其流动性，不但许可而且要求野战军的大踏步的进退。积极方面，为了陷敌于不利而利于我之作战，常常要求敌人在运动中，并要求有利于我之许多条件，例如有利的地形、好打的敌情、能封锁消息的居民、敌人的疲劳和不意等。这就要求敌人的前进，虽暂时地丧失部分土地而不惜。因为暂时地部分地丧失土地，是全部地永久地保存土地和恢复土地的代价。消极方面，凡被迫处于不利地位，根本上危及军力的保存时，应该勇敢地退却，以便保存军力，在新的时机中再行打击敌人。拚命主义者不知此理，明明已处于确定了的不利情况，还要争一城一地的得失，结果不但城和地俱失，军力也不能保存。我们历来主张"诱敌深入"，就是因为这是战略防御中弱军对强军作战的最有效的军事政策。

（九五）抗日战争的作战形式中，主要的是运动战，其次就要算游击战了。我们说，整个战争中，运动战是主要的，游击战是辅助的，说的是解决战争的命运，主要是依靠正规战，尤其是其中的运动战，游击战不能担负这种解决战争命运的主要的责任。但这不是说：游击战在抗日战争中的战略地位不重要。游击战在整个抗日战争中的战略地位，仅仅次于运动战，因为没有游击战的辅助，也就不能战胜敌人。这样说，是包括了游击战向运动战发展这一个战略任务在内的。长期的残酷的战争中间，游击战不停止于原来地位，它将把自己提高到运动战。这样，游击战的战略作用就有两方面：一是辅助正规战，一是把自己也变为正规战。至于就游击战在中国抗日战争中的空前广大和空前持久的意义说来，它的战略地位是更加不能轻视的了。因此，在中国，游击战的本身，不只有战术问题，还有它的特殊的战略问题。这个问题，我在《抗日游击战争的战略问题》一文里面已经说到了。前面说过，抗日战争三个战略阶段的作战形式，第一阶段，运动战是主要的，游击战和阵地战是辅助的。第二阶段，则游击战将升到主要地位，而以运动战和阵地战辅助之。第三阶段，运动战再升为主要形式，而辅之以阵地战和游击战。但这个第三阶段的运动战，已不全是由原来的正规军负担，而将由原来的游击军从游击战提高到运动战去担负其一部分，也许是相当重要的一部分。从三个阶段来看，中国抗日战争中的游击战，决不是可有可无的。它将在人类战争史上演出空前伟大的一幕。为此缘故，在全国的数百万正规军中间，至少指定数十万人，分散于所有一切敌占地区，发动和配合民众武装，从事游击战争，是完全必要的。被指定的军队，要自觉地负担这种神圣任务，不要以为少打大仗，一时显得不像民族英雄，降低了资格，这种想法是错误的。游击战争没有正规战争那样迅速的成效和显赫的名声，但是"路遥知马力，事久见人心"，在长期和残酷的战争中，游击战争将表现其很大的威力，实在是非同小可的事业。并且正规军分散作游击战，集合起来又可作运动战，八路军就是这样做的。八路军的方针是："基本的是游击战，但不放松有利条件下的运动战。"这个方针是完全正确的，反对这个方针的人们的观点是不正确的。

（九六）防御的和攻击的阵地战，在中国今天的技术条件下，一般都不能执行，这也就是我们表现弱的地方。再则敌人又利用中国土地广大一点，回避我们的阵地设施。因此阵地战就不能用为重要手段，更不待说用为主要手段。然而在战争的第一第二两阶段中，包括于运动战范围，而在战役作战上起其辅助作用的局部的阵地战，是可能的和必要的。为着节节抵抗以求消耗敌人和争取余裕时间之目的，而采取半阵地性的所谓"运动性的防御"，更是属于运动战的必要部分。中国须努力增加新式武器，以便在战略反攻阶段中能够充分地执行阵地攻击的任务。战略反攻阶段，无疑地将提高阵地战的地位，因为那时敌人将坚守阵地，没有我之有力的阵地攻击以配合运动战，将不能达到收复失地之目的。虽然如此，第三阶段中，我们仍须力争以运动战为战争的主要形式。因为战争的领导艺术和人的活跃性，临到像第一次世界大战的中期以后西欧地区那样的阵地战，就死了一大半。然而在广大版图的中国境内作战，在相当长的时间内中国方面又还保存着技术贫弱这种情况，"把战争从壕沟里解放"的事，就自然出现。就在第三阶段，中国技术条件虽已增进，但仍不见得能够超过敌人，这样也就被逼着非努力讲求高度的运动战，不能达到最后胜利之目的。这样，整个抗日战争中，中国将不会以阵地战为主要形式，主要和重要的形式是运动战和游击战。在这些战争形式中，战争的领导艺术和人的活跃性能够得到充分地发挥的机会，这又是我们不幸中的幸事啊！

消耗战，歼灭战

（九七）前头说过，战争本质即战争目的，是保存自己，消灭敌人。然而达此目的的战争形式，有运动战、阵地战、游击战三种，实现时的效果就有程度的不同，因而一般地有所谓消耗战和歼灭战之别。

（九八）我们首先可以说，抗日战争是消耗战，同时又是歼灭战。为什么？敌之强的因素尚在发挥，战略上的优势和主动依然存在，没有战役和战斗的歼灭战，就不能有效地迅速地减杀其强的因素，破坏其优势和主动。我之弱的因素也依然存在，战略上的劣势和被动还未脱离，为了争取时间，加强国内国际条件，改变自己的不利状态，没有战役和战斗的歼灭战，也不能成功。因此，战役的歼灭战是达到战略的消耗战之目的的手段。在这点上说，歼灭战就是消耗战。中国之能够进行持久战，用歼灭达到消耗是主要的手段。（九九）但达到战略消耗目的的，还有战役的消耗战。大抵运动战是执行歼灭任务的，阵地战是执行消耗任务的，游击战是执行消耗任务同时又执行歼灭任务的，三者互有区别。在这点上说，歼灭战不同于消耗战。战役的消耗战，是辅助的，但也是持久作战所需要的。

（一〇〇）从理论上和需要上说来，中国在防御阶段中，应该利用运动战之主要的歼灭性，游击战之部分的歼灭性，加上辅助性质的阵地战之主要的消耗性和游击战之部分的消耗性，用以达到大量消耗敌人的战略目的。在相持阶段中，继续利用游击战和运动战的歼灭性和消耗性，再行大量地消耗敌人。所有这些，都是为了使战局持久，逐渐地转变敌我形势，准备反攻的条件。战略反攻时，继续用歼灭达到消耗，以便最后地驱逐敌人。

（一〇一）但是在事实上，十个月的经验是，许多甚至多数的运动战战役，打成了消耗战；游击战之应有的歼灭作用，在某些地区，也还未提到应有的程度。这种情况的好处是，无论如何我们总算消耗了敌人，对于持久作战和最后胜利有其意义，我们的血不是白流的。然而缺点是：一则消耗敌人的不足；二则我们自己不免消耗的较多，缴获的较少。虽然应该承认这种情况的客观原因，即敌我技术和兵员教养程度的不同，然而在理论上和实际上，无论如何也应该提倡主力军在一切有利场合努力地执行歼灭战。游击队虽然为了执行许多具体任务，例如破坏和扰乱等，不能不进行单纯的消耗战，然而仍须提倡并努力实行在战役和战斗之一切有利场合的歼灭性的作战，以达既能大量消耗敌人又能大量补充自己之目的。

（一○二）外线的速决的进攻战之所谓外线，所谓速决，所谓进攻，与乎运动战之所谓运动，在战斗形式上，主要地就是采用包围和迂回战术，因而便须集中优势兵力。所以，集中兵力，采用包围迂回战术，是实施运动战即外线的速决的进攻战之必要条件。然而一切这些，都是为着歼灭敌人之目的。

（一○三）日本军队的长处，不但在其武器，还在其官兵的教养——其组织性，其因过去没有打过败仗而形成的自信心，其对天皇和对鬼神的迷信，其骄慢自尊，其对中国人的轻视等等特点；这是日本军阀多年的武断教育和日本的民族习惯造成的。我军对之杀伤甚多、俘虏甚少的现象，主要原因在此。这一点，过去许多人是估计不足的。这种东西的破坏，需要一个长的过程。首先需要我们重视这一特点，然后耐心地有计划地从政治上、国际宣传上、日本人民运动上多方面地向着这一点进行工作；而军事上的歼灭战，也是方法之一。在这里，悲观主义者可以据之引向亡国论，消极的军事家又可以据之反对歼灭战。我们则相反，我们认为日本军队的这种长处是可以破坏的，并且已在开始破坏中。破坏的方法，主要的是政治上的争取。对于日本士兵，不是侮辱其自尊心，而是了解和顺导他们的这种自尊心，从宽待俘虏的方法，引导他们了解日本统治者之反人民的侵略主义。另一方面，则是在他们面前表示中国军队和中国人民不可屈服的精神和英勇顽强的战斗力，这就是给以歼灭战的打击。在作战上讲，十个月的经验证明歼灭是可能的，平型关、台儿庄等战役就是明证。日本军心已在开始动摇，士兵不了解战争目的，陷于中国军队和中国人民的包围中，冲锋的勇气远弱于中国兵等等，都是有利于我之进行歼灭战的客观的条件，这些条件并将随着战争之持久而日益发展起来。在以歼灭战破坏敌军的气焰这一点上讲，歼灭又是缩短战争过程提早解放日本士兵和日本人民的条件之一。世界上只有猫和猫做朋友的事，而没有猫和老鼠做朋友的事。

（一○四）另一方面，应该承认在技术和兵员教养的程度上，现时我们不及敌人。因而最高限度的歼灭，例如全部或大部俘获的事，在许多场合特别是在平原地带的战斗中，是困难的。速胜论者在这点上面的过分要求，也属不对。抗日战争的正确要求应该是：尽可能的歼灭战。在一切有利的场合，每战集中优势兵力，采用包围迂回战术——不能包围其全部也包围其一部，不能俘获所包围之全部也俘获所包围之一部，不能俘获所包围之一部也大量杀伤所包围之一部。而在一切不利于执行歼灭战的场合，则执行消耗战。对于前者，用集中兵力的原则；对于后者，用分散兵力的原则。在战役的指挥关系上，对于前者，用集中指挥的原则；对于后者，用分散指挥的原则。这些，就是抗日战争战场作战的基本方针。

乘敌之隙的可能性

（一○五）关于敌之可胜，就是在敌人的指挥方面也有其基础。自古无不犯错误的将军，敌人之有岔子可寻，正如我们自己也难免出岔子，乘敌之隙的可能性是存在的。从战略和战役上说来，敌人在十个月侵略战争中，已经犯了许多错误。计其大者有五。一是逐渐增加兵力。这是由于敌人对中国估计不足而来的，也有他自己兵力不足的原因。敌人一向看不起我们，东四省[40]得了便宜之后，加之以冀东、察北的占领，这些都算作敌人的战略侦察。他们得来的结论是：一盘散沙。据此以为中国不值一打，而定出所谓"速决"的计划，少少出点兵力，企图吓溃我们。十个月来，中国这样大的团结和这样大的抵抗力，他们是没有料到的，他们把中国已处于进步时代，中国已存在着先进的党派、先进的军队和先进的人民这一点忘掉了。及至不行，就逐渐增兵，由十几个师团一次又一次地增至三十个。再要前进，非再增不可。但由于同苏联对立，又由于人财先天不足，所以日本的最大的出兵数和最后的进攻点都不得不受一定的限制。二是没有主攻方向。台儿庄战役以前，敌在华中、华北大体上是平分兵力的，两方内部又各自平分。例如华北，在津浦、平汉、同蒲三路平分兵力，每路伤亡了一些，占领地驻守了一些，再前进就没有兵了。台儿庄败仗后，总结了教训，把主力集中徐州方向，这个错误算是暂时地改了一下。三是没有战略协同。敌之华中、华北两集团中，每一集团内部是大体协同的，但两集团间则很不协同。津浦南段打小蚌埠时，北段不动；北段打台儿庄时，南段不动。两处都触了霉头之后，于是陆军大臣来巡视了，参谋总长来指挥了，算是暂时地协调了一下。日本地主资产阶级和军阀内部存在着颇为严重的矛盾，这种矛盾正在向前发展着，战争的不协同是其具体表现之一。四是失去战略时机。这点显着地表现在南京、太原两地占领后的停顿，主要的是因为兵力不足，没有战略追击队。五是包围多歼灭少。台儿庄战役以前，上海、南京、沧州、保定、南口、忻口、临汾诸役，击破者多，俘获者少，表现其指挥的笨拙。这五个——逐渐增加兵力，没有主攻方向，没有战略协同，失去时机，包围多歼灭少，是台儿庄战役以前日本指挥的不行之点。台儿庄战役以后，虽已改了一些，然根据其兵力不足和内部矛盾诸因素，求不重犯错误是不可能的。且得之于此者，又失之于彼。例如，将华北兵力集中于徐州，华北占领地就出了大空隙，给予游击战争以放手发展的机会。以上是敌人自己弄错，不是我们使之错的。我们方面，尚可有意地制造敌之错误，即用自己聪明而有效的动作，在有组织的民众掩护之下，造成敌人错觉，调动敌人就我范围，例如声东击西之类，这件事的可能性前面已经说过了。所有这些，都说明：我之战争胜利又可在敌之指挥上面找到某种根源。虽然我们不应把这点作为我之战略计划的重要基础，相反，我之计划宁可放在敌人少犯错误的假定上，才是可靠的做法。而且我乘敌隙，敌也可以乘我之隙，少授敌以可寻之隙，又是我们指挥方面的任务。然而敌之指挥错误，是事实上已经存在过，并且还要发生的，又可因我之努力制造出来的，都足供我之利用，抗日将军们应该极力地捉住它。敌人的战略战役指挥许多不行，但其战斗指挥，即部队战术和小兵团战术，却颇有高明之处，这一点我们应该向他学习。

抗日战争中的决战问题

（一○六）抗日战争中的决战问题应分为三类：一切有把握的战役和战斗应坚决地进行决战，一切无把握的战役和战斗应避免决战，赌国家命运的战略决战应根本避免。抗日战争不同于其它许多战争的特点，又表现在这个决战问题上。在第一第二阶段，敌强我弱，敌之要求在于我集中主力与之决战。我之要求则相反，在选择有利条件，集中优势兵力，与之作有把握的战役和战斗上的决战，例如平型关、台儿庄以及许多的其它战斗；而避免在不利条件下的无把握的决战，例如彰德等地战役所采的方针。拚国家命运的战略的决战则根本不干，例如最近之徐州撤退。这样就破坏了敌之"速决"计划，不得不跟了我们干持久战。这种方针，在领土狭小的国家是做不到的，在政治太落后了的国家也难做到。我们是大国，又处进步时代，这点是可以做到的。如果避免了战略的决战，"留得青山在，不愁没柴烧"，虽然丧失若干土地，还有广大的回旋余地，可以促进并等候国内的进步、国际的增援和敌人的内溃，这是抗日战争的上策。急性病的速胜论者熬不过持久战的艰难路程，企图速胜，一到形势稍为好转，就吹起了战略决战的声浪，如果照了干去，整个的抗战要吃大亏，持久战为之葬送，恰恰中了敌人的毒计，实在是下策。不决战就须放弃土地，这是没有疑问的，在无可避免的情况下（也仅仅是在这种情况下），只好勇敢地放弃。情况到了这种时候，丝毫也不应留恋，这是以土地换时间的正确的政策。历史上，俄国以避免决战，执行了勇敢的退却，战胜了威震一时的拿破仑[41]。中国现在也应这样干。

（一○七）不怕人家骂"不抵抗"吗？不怕的。根本不战，与敌妥协，这是不抵抗主义，不但应该骂，而且完全不许可的。坚决抗战，但为避开敌人毒计，不使我军主力丧于敌人一击之下，影响到抗战的继续，一句话，避免亡国，是完全必需的。在这上面发生怀疑，是战争问题上的近视眼，结果一定和亡国论者走到一伙去。我们曾经批评了所谓"有进无退"的拚命主义，就是因为这种拚命主义如果成为一般的风气，其结果就有使抗战不能继续，最后引向亡国的危险。

（一○八）我们主张一切有利条件下的决战，不论是战斗的和大小战役的，在这上面不容许任何的消极。给敌以歼灭和给敌以消耗，只有这种决战才能达到目的，每个抗日军人均须坚决地去做。为此目的，部分的相当大量的牺牲是必要的，避免任何牺牲的观点是懦夫和恐日病患者的观点，必须给以坚决的反对。李服膺、韩复榘等逃跑主义者的被杀，是杀得对的。在战争中提倡勇敢牺牲英勇向前的精神和动作，是在正确的作战计划下绝对必要的东西，是同持久战和最后胜利不能分离的。我们曾经严厉地指斥了所谓"有退无进"的逃跑主义，拥护严格纪律的执行，就是因为只有这种在正确计划下的英勇决战，才能战胜强敌；而逃跑主义，则是亡国论的直接支持者。

（一〇九）英勇战斗于前，又放弃土地于后，不是自相矛盾吗？这些英勇战斗者的血，不是白流了吗？这是非常不妥当的发问。吃饭于前，又拉屎于后，不是白吃了吗？睡觉于前，又起床于后，不是白睡了吗？可不可以这样提出问题呢？我想是不可以的。吃饭就一直吃下去，睡觉就一直睡下去，英勇战斗就一直打到鸭绿江，这是主观主义和形式主义的幻想，在实际生活里是不存在的。谁人不知，为争取时间和准备反攻而流血战斗，某些土地虽仍不免于放弃，时间却争取了，给敌以歼灭和给敌以消耗的目的却达到了，自己的战斗经验却取得了，没有起来的人民却起来了，国际地位却增长了。这种血是白流的吗？一点也不是白流的。放弃土地是为了保存军力，也正是为了保存土地；因为如不在不利条件下放弃部分的土地，盲目地举行绝无把握的决战，结果丧失军力之后，必随之以丧失全部的土地，更说不到什么恢复失地了。资本家做生意要有本钱，全部破产之后，就不算什么资本家。赌汉也要赌本，孤注一掷，不幸不中，就无从再赌。事物是往返曲折的，不是径情直遂的，战争也是一样，只有形式主义者想不通这个道理。

（一一〇）我想，即在战略反攻阶段的决战亦然。那时虽然敌处劣势，我处优势，然而仍适用"执行有利决战，避免不利决战"的原则，直至打到鸭绿江边，都是如此。这样，我可始终立于主动，一切敌人的"挑战书"，旁人的"激将法"，都应束之高阁，置之不理，丝毫也不为其所动。抗日将军们要有这样的坚定性，才算是勇敢而明智的将军。那些"一触即跳"的人们，是不足以语此的。第一阶段我处于某种程度的战略被动，然在一切战役上也应是主动的，尔后任何阶段都应是主动。我们是持久论和最后胜利论者，不是赌汉们那样的孤注一掷论者。

兵民是胜利之本

（———）日本帝国主义处在革命的中国面前，是决不放松其进攻和镇压的，它的帝国主义本质规定了这一点。中国不抵抗，日本就不费一弹安然占领中国，东四省的丧失，就是前例。中国若抵抗，日本就向着这种抵抗力压迫，直至它的压力无法超过中国的抵抗力才停止，这是必然的规律。日本地主资产阶级的野心是很大的，为了南攻南洋群岛，北攻西伯利亚起见，采取中间突破的方针，先打中国。那些认为日本将在占领华北、江浙一带以后适可而止的人，完全没有看到发展到了新阶段迫近了死亡界线的日本帝国主义，已经和历史上的日本不相同了。我们说，日本的出兵数和进攻点有一定的限制，是说：在日本一方面，在其力量基础上，为了还要举行别方面的进攻并防御另一方面的敌人，只能拿出一定程度的力量打中国打到它力所能及的限度为止；在中国一方面，又表现了自己的进步和顽强的抵抗力，不能设想只有日本猛攻，中国没有必要的抵抗力。日本不能占领全中国，然而在它一切力所能及的地区，它将不遗余力地镇压中国的反抗，直至日本的内外条件使日本帝国主义发生了进入坟墓的直接危机之前，它是不会停止这种镇压的。日本国内的政治只有两个出路：或者整个当权阶级迅速崩溃，政权交给人民，战争因而结束，但暂时无此可能；或者地主资产阶级日益法西斯化，把战争支持到自己崩溃的一天，日本走的正是这条路。除此没有第三条路。那些希望日本资产阶级中和派出来停止战争的，仅仅是一种幻想而已。日本的资产阶级中和派，已经作了地主和金融寡头的俘虏，这是多年来日本政治的实际。日本打了中国之后，如果中国的抗战还没有给日本以致命的打击，日本还有足够力量的话，它一定还要打南洋或西伯利亚，甚或两处都打。欧洲战争一起来，它就会干这一手；日本统治者的如意算盘是打得非常之大的。当然存在这种可能：由于苏联的强大，由于日本在中国战争中的大大削弱，它不得不停止进攻西伯利亚的原来计划，而对之采取根本的守势。然而在出现了这种情形之时，不是日本进攻中国的放松，反而是它进攻中国的加紧，因为那时它只剩下了向弱者吞剥的一条路。那时中国的坚持抗战、坚持统一战线和坚持持久战的任务，就更加显得严重，更加不能丝毫懈气。

（一一二）在这种情况下，中国制胜日本的主要条件，是全国的团结和各方面较之过去有十百倍的进步。中国已处于进步的时代，并已有了伟大的团结，但是目前的程度还非常之不够。日本占地如此之广，一方面由于日本之强，一方面则由于中国之弱；而这种弱，完全是百年来尤其是近十年来各种历史错误积累下来的结果，使得中国的进步因素限制在今天的状态。现在要战胜这样一个强敌，非有长期的广大的努力是不可能的。应该努力的事情很多，我这里只说最根本的两方面：军队和人民的进步。

（一一三）革新军制离不了现代化，把技术条件增强起来，没有这一点，是不能把敌人赶过鸭绿江的。军队的使用需要进步的灵活的战略战术，没有这一点，也是不能胜利的。然而军队的基础在士兵，没有进步的政治精神贯注于军队之中，没有进步的政治工作去执行这种贯注，就不能达到真正的官长和士兵的一致，就不能激发官兵最大限度的抗战热忱，一切技术和战术就不能得着最好的基础去发挥它们应有的效力。我们说日本技术条件虽优，但它终必失败，除了我们给以歼灭和消耗的打击外，就是它的军心终必随着我们的打击而动摇，武器和兵员结合不稳。我们相反，抗日战争的政治目的是官兵一致的。在这上面，就有了一切抗日军队的政治工作的基础。军队应实行一定限度的民主化，主要地是废除封建主义的打骂制度和官兵生活同甘苦。这样一来，官兵一致的目的就达到了，军队就增加了绝大的战斗力，长期的残酷的战争就不患不能支持。

（一一四）战争的伟力之最深厚的根源，存在于民众之中。日本敢于欺负我们，主要的原因在于中国民众的无组织状态。克服了这一缺点，就把日本侵略者置于我们数万万站起来了的人民之前，使它像一匹野牛冲入火阵，我们一声唤也要把它吓一大跳，这匹野牛就非烧死不可。我们方面，军队须有源源不绝的补充，现在下面胡干的"捉兵法"、"买兵法"[42]，亟须禁止，改为广泛的热烈的政治动员，这样，要几百万人当兵都是容易的。抗日的财源十分困难，动员了民众，则财政也不成问题，岂有如此广土众民的国家而患财穷之理？军队须和民众打成一片，使军队在民众眼睛中看成是自己的军队，这个军队便无敌于天下，个把日本帝国主义是不够打的。

（一一五）很多人对于官兵关系、军民关系弄不好，以为是方法不对，我总告诉他们是根本态度（或根本宗旨）问题，这态度就是尊重士兵和尊重人民。从这态度出发，于是有各种的政策、方法、方式。离了这态度，政策、方法、方式也一定是错的，官兵之间、军民之间的关系便决然弄不好。军队政治工作的三大原则：第一是官兵一致，第二是军民一致，第三是瓦解敌军。这些原则要实行有效，都须从尊重士兵、尊重人民和尊重已经放下武器的敌军俘虏的人格这种根本态度出发。那些认为不是根本态度问题而是技术问题的人，实在是想错了，应该加以改正才对。

（一一六）当此保卫武汉等地成为紧急任务之时，发动全军全民的全部积极性来支持战争，是十分严重的任务。保卫武汉等地的任务，毫无疑义必须认真地提出和执行。然而究竟能否确定地保卫不失，不决定于主观的愿望，而决定于具体的条件。政治上动员全军全民起来奋斗，是最重要的具体的条件之一。不努力于争取一切必要的条件，甚至必要条件有一不备，势必重蹈南京等地失陷之覆辙。中国的马德里在什么地方，看什么地方具备马德里的条件。过去是没有过一个马德里的，今后应该争取几个，然而全看条件如何。条件中的最基本条件，是全军全民的广大的政治动员。

（一一七）在一切工作中，应该坚持抗日民族统一战线的总方针。因为只有这种方针才能坚持抗战，坚持持久战，才能普遍地深入地改善官兵关系、军民关系，才能发动全军全民的全部积极性，为保卫一切未失地区、恢复一切已失地区而战，才能争取最后胜利。

（——八）这个政治上动员军民的问题，实在太重要了。我们之所以不惜反反复复地说到这一点，实在是没有这一点就没有胜利。没有许多别的必要的东西固然也没有胜利，然而这是胜利的最基本的条件。抗日民族统一战线是全军全民的统一战线，决不仅仅是几个党派的党部和党员们的统一战线；动员全军全民参加统一战线，才是发起抗日民族统一战线的根本目的。

结 论

（——九）结论是什么呢？结论就是："在什么条件下，中国能战胜并消灭日本帝国主义的实力呢？要有三个条件：第一是中国抗日统一战线的完成；第二是国际抗日统一战线的完成；第三是日本国内人民和日本殖民地人民的革命运动的兴起。就中国人民的立场来说，三个条件中，中国人民的大联合是主要的。""这个战争要延长多久呢？要看中国抗日统一战线的实力和中日两国其它许多决定的因素如何而定。""如果这些条件不能很快实现，战争就要延长。但结果还是一样，日本必败，中国必胜。只是牺牲会大，要经过一个很痛苦的时期。""我们的战略方针，应该是使用我们的主力在很长的变动不定的战线上作战。中国军队要胜利，必须在广阔的战场上进行高度的运动战。""除了调动有训练的军队进行运动战之外，还要在农民中组织很多的游击队。""在战争的过程中……使中国军队的装备逐渐加强起来。因此，中国能够在战争的后期从事阵地战，对于日本的占领地进行阵地的攻击。这样，日本在中国抗战的长期消耗下，它的经济行将崩溃；在无数战争的消磨中，它的士气行将颓靡。中国方面，则抗战的潜伏力一天一天地奔腾高涨，大批的革命民众不断地倾注到前线去，为自由而战争。所有这些因素和其它的因素配合起来，就使我们能够对日本占领地的堡垒和根据地，作最后的致命的攻击，驱逐日本侵略军出中国。"（一九三六年七月与斯诺谈话）"中国的政治形势从此开始了一个新阶段，……这一阶段的最中心的任务是：动员一切力量争取抗战的胜利。""争取抗战胜利的中心关键，在使已经发动的抗战发展为全面的全民族的抗战。只有这种全面的全民族的抗战，才能使抗战得到最后的胜利。""由于当前的抗战还存在着严重的弱点，所以在今后的抗战过程中，可能发生许多挫败、退却，内部的分化、叛变，暂时和局部的妥协等不利的情况。因此，应该看到这一抗战是艰苦的持久战。但我们相信，已经发动的抗战，必将因为我党和全国人民的努力，冲破一切障碍物而继续地前进和发展。"（一九三七年八月《中共中央关于目前形势与党的任务的决定》）这些就是结论。亡国论者看敌人如神物，看自己如草芥，速胜论者看敌人如草芥，看自己如神物，这些都是错误的。我们的意见相反：抗日战争是持久战，最后胜利是中国的——这就是我们的结论。

（一二○）我的讲演至此为止。伟大的抗日战争正在开展，很多人希望总结经验，以便争取全部的胜利。我所说的，只是十个月经验中的一般的东西，也算一个总结吧。这个问题值得引起广大的注意和讨论，我所说的只是一个概论，希望诸位研究讨论，给以指正和补充。

注释

[1] 见本卷《反对日本进攻的方针、办法和前途》注〔1〕。

[2] 见本书第一卷《中国革命战争的战略问题》注〔45〕。

[3] 这种亡国论是国民党内部分领导人的意见。他们是不愿意抗日的，后来抗日是被迫的。卢沟桥事变以后，蒋介石一派参加抗日了，汪精卫一派就代表了亡国论，并准备投降日本，后来果然投降了。但是亡国论思想不但是在国民党内存在着，在某些中层社会中甚至在一部分落后的劳动人民中也曾经发生影响。这是因为国民党政府腐败无能，在抗日战争中节节失败，而日军则长驱直进，在战争的第一年中就侵占了华北和华中的大片土地，因而在一部分落后的人民中产生了严重的悲观情绪。

[4] 以上这些意见，都是共产党内的。在抗日战争的头半年内，党内存在着一种轻敌的倾向，认为日本不值一打。其根据并不是因为他们感觉自己的力量很大，他们知道共产党领导的军队和民众的有组织的力量在当时还是很小的；而是因为国民党抗日了，他们感觉国民党有很大的力量，可以有效地打击日本。他们只看见国民党暂时抗日的一面，忘记了国民党反动和腐败的一面，因而造成了错误的估计。

[5] 这是蒋介石等人的意见。蒋介石国民党既已被迫抗战，他们就一心希望外国的迅速援助，不相信自己的力量，更不相信人民的力量。

[6] 一九三八年三月下旬至四月上旬，中国军队和日本侵略军在台儿庄（今属山东省枣庄市）一带进行过一次会战。在这次会战中，中国军队击败日军第五、第十两个精锐师团，取得了会战的胜利。

[7] 徐州战役是中国军队同日本侵略军在以徐州为中心的广大地区进行的一次战役。从一九三七年十二月起，华北、华中的日军分南北两线沿津浦铁路和台潍（台儿庄至潍县）公路进犯徐州外围地区。一九三八年四月上旬，中国军队在取得台儿庄会战的胜利后，继续

向鲁南增兵，在徐州附近集结了约六十万的兵力；而日军在台儿庄遭到挫败以后，从四月上旬开始调集南北两线兵力二十多万人，对徐州进行迂回包围。中国军队在日军夹击和包围下，分路向豫皖边突围。五月十九日，徐州被日军占领。

[8] 这是当时《大公报》在一九三八年四月二十五日和二十六日社评中提出的意见。他们从一种侥幸心理出发，希望用几个台儿庄一类的胜仗就能打败日本，免得在持久战中动员人民力量，危及自己阶级的安全。当时国民党统治集团内普遍有这种侥幸心理。

[9] 参见本书第一卷《论反对日本帝国主义的策略》注〔33〕。抗日战争时期，托派在宣传上主张抗日，但是攻击中国共产党的抗日民族统一战线政策。把托派与汉奸相提并论，是由于当时在共产国际内流行着中国托派与日本帝国主义间谍组织有关的错误论断所造成的。

[10] 见本书第一卷《论反对日本帝国主义的策略》注〔35〕。

[11] 见本书第一卷《论反对日本帝国主义的策略》注〔36〕。

[12] 戊戌维新也称戊戌变法，是一八九八年（戊戌年）发生的维新运动。当时，中国面临被帝国主义列强瓜分的严重危机。康有为、梁启超、谭嗣同等人，在清朝光绪皇帝的支持下，企图通过自上而下的变法维新，逐步地在中国推行地主阶级和资产阶级联合统治的君主立宪制度，发展民族资本主义，以挽救民族危亡。但是，这个运动缺乏人民群众的基础，又遭到以慈禧太后为首的顽固派的坚决反对。变法三个多月以后，慈禧太后发动政变，幽禁光绪皇帝，杀害谭嗣同等六人，变法遭到失败。

[13] 见本书第一卷《湖南农民运动考察报告》注〔3〕。

[14] 一九三八年一月十六日，日本近卫内阁发表声明，宣布以武力灭亡中国的方针；同时宣称由于国民党政府仍在"策划抗战"，日本政府决定在中国扶植新的傀儡政权，"今后将不以国民政府为对手"。

[15] 这里主要是指美国。自一九三七年到一九四○年，美国每年输入日本的物资占日本全部进口额的三分之一以上，其中战争物资占一半以上。

[16] 指英、美、法等帝国主义国家的政府。

[17] 张伯伦（一八六九——一九四○），英国保守党领袖。一九三七年至一九四○年任英国首相。他主张迁就德、意、日法西斯对中国、埃塞俄比亚、西班牙、奥地利和捷克斯洛伐克等国家的侵略，实行妥协政策。

[18] 毛泽东在这里所预言的抗日战争相持阶段中中国方面可能的向上变化，在中国共产党领导下的抗日根据地是完全实现了。在国民党统治区，则因为以蒋介石为首的统治集团消极抗日、积极反共反人民，不但没有向上变化，反而向下变化了。因为这样，也激起了广大人民的反抗和觉悟。参见本书第三卷《论联合政府》第三部分关于这一切事实的分析。

[19] 这个比喻里所引用的神话故事，见明朝吴承恩所著的《西游记》第七回。这个神话故事说，孙悟空本是个猴子，他能够一个筋斗翻十万八千里，但是，他站在如来佛的手心上尽力翻筋斗，总是翻不出去。如来佛翻掌一扑，将五个手指化作五行山，把他压住。

[20] 一九三五年八月，季米特洛夫在共产国际第七次代表大会上所作的报告中说："法西斯是肆无忌惮的沙文主义和侵略战争。"一九三

七年七月，他又发表了题为《法西斯主义就是战争》的论文。

[21] 参见列宁《第二国际的破产》和《社会主义与战争》（《列宁全集》第26卷，人民出版社1988年版，第235、327页）。

[22] 见本卷《为动员一切力量争取抗战胜利而斗争》。

[23] 见本卷《陕甘宁边区政府、第八路军后方留守处布告》注〔3〕。

[24] 见《孙子·谋攻》。

[25] 城濮在今山东省鄄城县西南。公元前六三二年，晋楚两国大战于此。战争开始时，楚军占优势。晋军退却九十里，到达城濮一带，先选择楚军力量薄弱的右翼，给以严重的打击。然后，再集中优势兵力击溃了楚军的左翼。楚军终于大败而退。

[26] 见本书第一卷《中国革命战争的战略问题》注〔31〕。

[27] 公元前二〇四年，汉将韩信率部与赵王歇大战于井陉（在今河北省井陉县）。赵军号称二十万，数倍于汉军。韩信背水为阵，率军奋战；同时，遣兵袭占赵军防御薄弱的后方，使其腹背受敌，遂大破赵军。

[28] 见本书第一卷《中国革命战争的战略问题》注〔32〕。

[29] 见本书第一卷《中国革命战争的战略问题》注〔33〕。

[30] 见本书第一卷《中国革命战争的战略问题》注〔34〕。

[31] 见本书第一卷《中国革命战争的战略问题》注〔35〕。

[32] 见本书第一卷《中国革命战争的战略问题》注〔36〕。

[33] 十八世纪末十九世纪初，法国的拿破仑曾与英、普、奥、俄以及欧洲其它很多国家作战。在多次战争中，拿破仑的部队在数量上都不如他的敌人，但都得到了胜利。

[34] 公元三八三年，秦王苻坚出兵攻晋。他依仗优势兵力，非常轻视晋军。晋军打败了秦军的前锋，从水陆两路继续前进，隔淝水同秦军对峙。苻坚登寿阳城（今安徽省寿县）瞭望，见晋兵布阵严整，又望见八公山上的草木，以为都是晋兵，觉得是遇到了劲敌，开始有惧色。随后在淝水决战中，强大的秦军终于被晋军打败。

[35] 蒋介石、汪精卫等在一九二七年背叛革命以后，进行十年的反人民战争，同时又在国民党统治区实行法西斯统治。这就使得中国人民没有可能广泛地组织起来。这个历史错误是应该由蒋介石为首的国民党反动派负责的。

[36] 宋襄公是公元前七世纪春秋时代宋国的国君。公元前六三八年宋国与强大的楚国作战，宋兵已经排列成阵，楚兵正在渡河。宋国有一个官员认为楚兵多宋兵少，主张利用楚兵渡河未毕的时机出击。但宋襄公说：不可，因为君子不乘别人困难的时候去攻打人家。楚兵渡河以后，还未排列成阵，宋国官员又请求出击。宋襄公又说：不可，因为君子不攻击不成阵势的队伍。一直等到楚兵准备好了以后，宋襄公才下令出击。结果宋国大败，宋襄公自己也受了伤。

[37] 见《宋史·岳飞传》。

[38] 见《礼记·中庸》。

[39] 一九三七年日本侵略军在占领北平、天津以后，不久即分兵沿津

浦铁路南下，进攻山东省。多年统治山东的国民党军阀韩复榘不战而逃。在一九三七年十二月下旬至一九三八年一月上旬的十多天里，他就放弃了山东中部和西南部的大片国土，从济南一直逃到山东、河南的边境。

[40] 见本书第一卷《论反对日本帝国主义的策略》注〔5〕。

[41] 一八一二年，拿破仑以五十万大军进攻俄国。当时俄军只有二十万人左右。为了避免不利于自己的决战，俄军实行战略退却，一直到放弃和焚毁了莫斯科。拿破仑的军队在深入俄国国土以后，遭到了俄国广大军民的坚决反抗，陷于饥寒困苦、后路被切断、四面被包围的绝境，最后不得不从莫斯科撤退。这时，俄军乘机大举反攻，拿破仑军仅剩二万余人逃离俄国国境。

[42] 国民党政府扩军的一种办法，是派军警四处捉拿人民去当兵，捉来的兵用绳捆索绑，形同囚犯。略为有钱的人，就向国民党政府的官吏行贿，出钱买人代替。

www.ingramcontent.com/pod-product-compliance
Lightning Source LLC
Chambersburg PA
CBHW010939140726
47988CB00010B/3514